Experimental Surgical Models *in the* Laboratory Rat

Experimental Surgical Models *in the* Laboratory Rat

Edited by

Alfredo Rigalli

Verónica Elina Di Loreto

CRC Press
Taylor & Francis Group
Boca Raton London New York

CRC Press is an imprint of the
Taylor & Francis Group, an **informa** business

CRC Press
Taylor & Francis Group
6000 Broken Sound Parkway NW, Suite 300
Boca Raton, FL 33487-2742

Printed in the United States of America on acid-free paper
10 9 8 7 6 5 4 3 2 1

International Standard Book Number-13: 978-1-4200-9326-1 (Hardcover)

Library of Congress Cataloging-in-Publication Data

Experimental surgical models in the laboratory rat / edited by Alfredo Rigalli and Veronica Elina Di Loreto.
p. ; cm.
Includes bibliographical references and index.
ISBN-13: 978-1-4200-9326-1 (hardcover : alk. paper)
ISBN-10: 1-4200-9326-6 (hardcover : alk. paper)
1. Rats as laboratory animals. 2. Rats--Surgery. I. Rigalli, Alfredo, 1959- II. Di Loreto, Veronica Elina, 1968- III. Title.
[DNLM: 1. Disease Models, Animal. 2. Rats. 3. Surgical Procedures, Operative--methods. QY 60.R6 E64 2009]

SF407.R38E97 2009
616'.02733--dc22 2008052007

Visit the Taylor & Francis Web site at
http://www.taylorandfrancis.com

and the CRC Press Web site at
http://www.crcpress.com

Contents

SECTION X Additional Techniques

SECTION XI Miscellaneous

Preface

This book is the direct consequence of many years of work in the surgical theatre where assorted techniques were developed or reproduced in the study of pharmacokinetics and pharmacology of fluorine-containing compounds, diabetes, and pancreatitis. The experience and knowledge accumulated through decades of work and the start of my doctorate career in Biomedical Sciences, where surgical experimentation with rats is one of the disciplines, brought about the writing of this book, which actually began in 1997.

Since then, the compilation of surgical technique descriptions has been an ongoing endeavor and included the help of many associates. Contributors to this book are researchers, professors, and students from the Bone Biology and Mineral Metabolism Laboratory, School of Medicine, Rosario National University; from the Pharmacology Division, Biochemical and Pharmaceutical School, Rosario National University; from the Institute of Cardiovascular Pathophysiology, School of Medicine, Buenos Aires University; and from the Pharmacology Division, School of Veterinary, La Plata National University, all in Argentina. Contributor chapters are based on surgical models developed or reproduced in the course of their research projects.

The main objective of this book, *Experimental Surgical Models in the Laboratory Rat,* is to contribute to the postgraduate studies of researchers in the biomedical area and in the study of the mechanism of action and the efficacy of drugs in different pathologies.

The election of an experimental model is the crucial point in a research project. The knowledge of where this model can be used, how it can be done, and when it can produce valuable results is not always clear for the researcher in the beginning. Once a model has been chosen, the aid of a teacher is important and can help reduce the optimization of the model, a process that usually takes months or years. The development of the methodology includes the correct selection of materials and procedures, the environmental conditions, the sex and strain of animals, the diet, and the personnel to take care of animals. The correct choice of all of these requirements will be crucial for obtaining reliable and reproducible results. This book provides important details that are not always included in the journals where the techniques are published. Sometimes small details are omitted, which can often make the difference between the success or failure of an experiment. It is important to notice that a failed experiment often implies the useless sacrifice of animals. An animal researcher has a moral obligation not to uselessly cause animal deaths without producing results.

Experimental Surgical Models in the Laboratory Rat is organized in sections, each of which contains a definite subject. All chapters are organized with a short introduction and the utility of the technique, the list of materials needed for performing the surgery, a step-by-step description of the surgery, and the precautions and experimental results obtained by the authors. Each chapter also includes detailed figures, which are complemented with sequential photographs of the surgery on a CD that accompanies this book. In the case that contingency plans are available, they are also described as well as the combination with other procedures.

The book contains a vast list of updated references where the theoretical bases of the models are described. Included as well are numerous journal articles where most of the results obtained with the models are included. Normal values of weight, food, and water consumption as well as some common biochemical parameters are included. These values come from an eumetabolic rat, which is no different than the recognized strains, such as Wistar and Sprague Dawley rats. Although there are a wide range of instruments available to assist in the surgery, a chapter explaining simple devices is also included.

Acknowledgments

To Dr. Rodolfo C. Puche, director of the Bone Biology and Mineral Metabolism Laboratory, Rosario, Argentina, who encourages young students in the development of animal models and is always enthusiastic about the results obtained from them.

To Dr. José Pellegrino, member of the Institute of Experimental Physiology, Rosario, Argentina, for helping with photographs and image processing.

To Dr. Lucas R. M. Brun, member of the Bone Biology and Mineral Metabolism Laboratory, Rosario, Argentina, for the design of images.

To Nicolás F. Rigalli for the preparation of files with photographs.

To Ariana Foresto and Maela Lupo, members of the Bone Biology and Mineral Metabolism Laboratory, Rosario, Argentina, for the grammatical revision of the manuscript.

To the members of the School of Medicine Vivarium, Rosario National University, Rosario, Argentina, for providing animals for the development of the models.

The Editors

Alfredo Rigalli, Ph.D., received his doctorate in biochemistry in 1990 from the Biochemical and Pharmacy School, Rosario National University. He has been a member of the Bone Biology and Mineral Metabolism Laboratory and assistant professor of biochemistry, Biochemistry Division, Physiology Department, School of Medicine, Rosario National University, in Rosario, Argentina since 1985.

Dr. Rigalli has also been a member of the National Scientific and Technical Research Council (CONICET) in Buenos Aires, Argentina, since 2001 and a member of the Research Council of Rosario National University (CIUNR) since 1990.

At the present time, Dr. Rigalli is professor of surgery in the rat, He teaches use of electrodes in the Biological Research Laboratory, bone biology, and mathematical modeling of biological processes in the Biomedical Doctorate School of Medicine and postgraduate courses in the Pharmacy and Biochemistry School and career specialization at the Dentistry School of Rosario National University.

Dr. Rigalli is a former president of the Argentine Association of Osteology and Bone Metabolism and his field of research includes pharmacokinetics, side and metabolic effects of fluorine containing compounds and bone metabolism. He is the author of several books and computer software for teaching biochemistry, mathematical modeling, and electrodes in graduate and postgraduate studies at Rosario National University.

Verónica Elina Di Loreto, Ph.D., has a degree in biochemistry and a Ph.D. in biomedical sciences. She has been a member of the Bone Biology and Mineral Metabolism Laboratory, Biochemistry Division, Physiology Department, School of Medicine, Rosario National University, Rosario, Argentina, since 1994.

Dr. Di Loreto is professor of surgery in the rat and bone metabolism in experimental diabetes, subjects of the Biomedical Doctorate, School of Medicine, Rosario National University, and assistant professor of biochemistry and biophysics in the School of Medicine.

Dr. Di Loreto is a former secretary and member of the Argentine Association of Osteology and Bone Metabolism as well as a member of the Argentine Society of Clinical Investigation. Her research interests include effects of fluoride compounds on bone and mineral metabolism and the effects of monofluorophosphate on pancreatitis evolution.

She has authored the book *Determination of Fluoride Concentration in Biological Samples* and written for several publications worldwide.

Contributors

Maria V. Arcidiacono, Ph.D. in physiopathology, pharmacology, clinical medicine and therapy of metabolic disease at the University of Milan, Italy, is a member of the Renal Division, Washington University, St. Louis, Missouri, USA.

María L. Brance, M.D., is on the faculty of the School of Medicine of Rosario National University, Rosario, Argentina. Dr. Brance is a professor of biochemistry and member of the Bone Biology and Mineral Metabolism Laboratory. She is also a member of the Rosario Biology Society and the Argentine Association of Osteology and Mineral Metabolism.

Anabel Brandoni, Ph.D., is on the faculty of the Pharmacy and Biochemist School of Rosario National University, Rosario, Argentina. She is Fellowship holder of the National Scientific and Technical Research Council (CONICET) and a member of the Argentine Society of Physiological Sciences.

Lucas R. M. Brun, M.D., is a member of the faculty of the School of Medicine of Rosario National University, Rosario, Argentina. Dr. Brun is a member of the Bone Biology and Mineral Metabolism Laboratory, the National Scientific and Technical Research Council (CONICET), the Rosario Biology Society, and Argentine Association of Osteology and Mineral Metabolism.

Bruno Buchholz, M.D., is a Fellowship holder of the National Scientific and Technologic National Agency, Buenos Aires, Argentina.

Verónica D´Annunzio, M.D., is a researcher for the Institute of Cardiovascular Pathophysiology, School of Medicine, University of Buenos Aires, Argentina.

Gisela Di Giusto has a Licenciate degree in biotechnology from Rosario National University, Rosario, Argentina. She is a member of the National Agency for the Promotion of Science and Technology (ANPCyT), Buenos Aires, Argentina.

Martín Donato, M.D., Ph.D., is a member of the National Scientific and Technical Research Council (CONICET) and a researcher for the Institute of Cardiovascular Pathophysiology, School of Medicine, University of Buenos Aires, Argentina.

Ariana Foresto is head of the foreign languages division of the Bone Biology and Mineral Metabolism Laboratory, Rosario National University, Rosario, Argentina.

Ricardo J. Gelpi, M.D., Ph.D., is a member of the National Scientific and Technical Research Council (CONICET), as well as head researcher for the Institute of Cardiovascular Pathophysiology, School of Medicine, University of Buenos Aires, Argentina.

Germán E. González, M.D., Ph.D., is a member of the National Scientific and Technical Research Council (CONICET). He is a researcher for the Institute of Cardiovascular Pathophysiology as well as an instructor of pathology at the School of Medicine, University of Buenos Aires, Argentina.

Gabriel A. Inchauspe, M.D., is a member of the Bone Biology and Mineral Metabolism Laboratory, and a faculty member of the School of Medicine. Rosario National University, Rosario, Argentina.

Laura Krieger, D.D., is a doctoral Fellow of the University of Buenos Aires. She is also a member of the International Association for Dental Research for the Argentine Association of Clinical Research.

María F. Landoni, Ph.D., is a professor of pharmacology at the Veterinary School, La Plata National University, Buenos Aires, Argentina. She also is a member of the National Council of Scientific and Technical Research (CONICET).

Mercedes Lombarte has a licenciate degree in biotechnology and is a member of the Bone Biology and Mineral Metabolism Laboratory at the School of Medicine, Rosario National University, Rosario, Argentina.

Maela Lupo holds a licenciate degree in biotechnology and is a member of the Bone Biology and Mineral Metabolism Laboratory at Rosario National University, Rosario, Argentina.

Inés Menoyo, Ph.D., is a member of the Bone Biology and Mineral Metabolism Laboratory and faculty member of the School of Medicine at the Rosario National University, Rosario, Argentina.

Hilda S. Moreno is a professional technical assistant for the National Council of Scientific and Technical Research (CONICET), Argentina.

Laura I. Pera, Ph.D., is a biochemist and a faculty member of the School of Medicine for the Rosario National University, Rosario, Argentina. She is a member of the Bone Biology and Mineral Metabolism Laboratory, the Rosario Biology Society, and the Argentine Association of Osteology and Mineral Metabolism.

Manuel Rodríguez, M.D., is a researcher for the Institute of Cardiovascular Pathophysiology and an instructor of Pathology at the School of Medicine, University of Buenos Aires, Argentina.

Adriana M. Torres, Ph.D., received a doctorate in biochemistry from Rosario National University, Rosario, Argentina. She is an assistant professor of Pharmacology and a member of the National Council of Scientific and Technical Research (CONICET) as well as a member of the Argentine Society of Clinical Investigation, Argentine Society of Physiological Sciences, and Rosario Biology Society.

Aneley A. Traverso, M.D., is a member of the Bone Biology and Mineral Metabolism Laboratory and a faculty member of the School of Medicine at Rosario National University, Rosario, Argentina.

Medical Students—*Fellows of the Bone Biology and Mineral Metabolism Laboratory*, School of Medicine, Rosario National University: Diego Holotte, Jésica Nipoti, María J. Pretini, Lucas F. de Candia, Valeria Dalmau, and Luis A. Ramirez; *Fellows of the Institute of Cardiovascular Pathophysiology*, School of Medicine, University of Buenos Aires: Cristina Lorenzo Carrión and Ignacio M. Seropian.

Abbreviations

APSB	alkaline phosphatase substrate buffer
BCIP	5-bromo-4-chloro-3-indolyl phosphate
BSA	bovine serum albumin
CNS	central nervous system
DMF	dimethylformamide
°C	degree Celsius
h	hour
hrGH	human recombining growth hormone
i.g.	intragastric
i.m.	intramuscular
i.p.	intraperitoneal
i.v.	intravenous
IU	international unit
l	liter
KRB	Krebs–Ringer buffer
KRBB	Krebs–Ringer bicarbonate buffer
MFP	sodium monofluorophosphate
ml	milliliter
mM	millimole/liter
NAD^+	nicotinamine adenin dinucleotide
OVX	ovariectomy
PAH	p-aminohippuric acid
p-NPP	p-nitrophenylphosphate
p.o.	oral
PTH	parathyroid hormone, parathormone
RIA	radioimmunoassay or radioimmunoanalysis
s.c.	subcutaneous
SD	standard deviation
SEM	standard error of the mean
STZ	streptozotocin
TPTX	thyroidparathyroidectomy
w/v	weight per volume
wt%	weight percent

Section I

Introduction

1 Bioethics and Animal Care

Lucas R. M. Brun and Verónica E. Di Loreto

INTRODUCTION

The use of animals with medical and biological objectives has been practiced for centuries. Areas of biomedical research, such as pharmacology, physiology, and toxicology, have based their progress in experiments carried out mainly in animal models or in the accurate observations of spontaneous phenomena in animals. Advances in human health conditions are, in part, the result of the knowledge of biological processes, which were first understood in animal models. Laboratory animals have also contributed in the development of vaccines, methodology for the diagnosis of different illnesses, identification of pharmacological target cells or molecules for new pharmaceutical products, organ transplantation, grafts, and much more.

Science has developed the model alternatives that do not involve animals, such as cell culture, mathematical models, computational simulation models, *in vitro* experiments with cell-free systems, etc. The question is: Is research with animals necessary? And it remains without a proper answer because it is influenced greatly by religion, human behavior, age, feelings and more. Although the question remains unanswered, there is agreement among scientists that even the more sophisticated technologies cannot reproduce the complexity and interactions among cells, molecules, tissues, and organs, which take place in a multicellular organism. However, these isolated systems provide information that cannot be obtained from intact animals because of their high complexity.

The results obtained in animal models provide invaluable information for the design of tests for new pharmacological products in human beings. The test of a new product in human beings must begin by experimenting on intact animals; this is apart from the tests in other models, such as cell culture. As a consequence, research with animals is an obligation in biomedical sciences. According to the Nuremberg Council, the tests carried out in human beings must be designed carefully and must take into consideration the results previously obtained in animals. The Declaration of Helsinki, which was adopted in 1964 by the World Medical Association, also indicates that the research in human beings must be designed based on results obtained with animals. Therefore, biomedical science is impossible without research with laboratory animals. However, the use of animals needs to be justified; it must follow national and international rules and interdisciplinary committees must evaluate the procedure where the experimental design is considered for both scientific and ethical principles.[1]

Along with the scientific knowledge gained from the use of laboratory animals in research, laws have been established that state principles for the use of animals as experimental models. Institutional animal care and use committees were created around the world to control the use of laboratory animals in research and for educational purposes. However, there are countries without this kind of legislation or where it is only now coming into existence.

The International Council for Laboratory Animal Science (ICLAS) is a nongovernmental organization for international cooperation in laboratory animal science.[2] This council's mission's is the advancing of human health by promoting the care and ethical use of laboratory animals. The first ICLAS meeting for harmonization of guidelines was held in 2004. One of the aims of this meeting was to obtain consensus from the most important organizations that deal with the care and ethical use of laboratory animals to develop guides of animal care. The last harmonization guidelines update was published in 2007.

There are important organizations around the world that regulate and produce guides for the use and management of laboratory animals, such as the Federation of European Laboratory Animal Science Associations (FELASA),[3] the American Association for Laboratory Animal Science (AALAS)[4], and the Canadian Council on Animal Care (CCAC).[5]

THE 3 Rs

The concept of the 3 Rs was established by Russell and Burch[6] in 1959. Since then, a number of changes have taken place in the use and management of laboratory animals in both research and educational objectives. The principle of the 3 Rs proposes the sensible and humanitarian use of animals in scientific work, and the aim is to guarantee the rational and respectful use of experimental animals, **reducing** the number of animals through the correct choice of genetic and environmental conditions, **replacing** the animal by another model when possible, and **refining** procedures in order to minimize stress and pain, but guaranteeing the validity of the results.

The alternatives of **reduction** describe methods to obtain valuable results from experiments carried out with few animals, such as the election of the correct animal model, pilot studies, correct design of the experiment, adequate statistical tests, sanitary quality, genetic and environmental qualities, etc. The correct and efficient bibliographic searching prevents the duplication of information and the realization of unnecessary experiments. The **replacing** alternatives refer to other methods that obtain the same results, but without involving animals, such as *in vitro* systems, cell culture, mathematical models, simulators, human materials, etc. The alternatives of **refining** add methods that alleviate or minimize pain, stress, anxiety and fear, and maintain the welfare of the animal. Refining relies on the knowledge and ability of trained personnel who deal with the animals and who have the capacity to detect pain and discomfort, use the appropriate analgesic and anesthetics, use mini- or noninvasive techniques, and use the correct choice of euthanasia. Refining in techniques produces better results and lower variability. For example, new anesthetics together with training in surgical techniques certainly reduce the number of deaths in the anesthetic procedure. In the same vein, knowledge about statistics and design of experiments contributes to the choosing of the adequate model and test without losing important information.

The researcher must act responsibly in order not to repeat experiments that have already been done. Only those experiments relevant and pertinent for scientific knowledge or the well-being of the community should be accepted.

The researcher must know when the results of the experiment are less important than the suffering and pain of the animals, and euthanasia must be performed even though the results at the end of the experiment are important. So, analgesia and euthanasia play an important role in the end point of an experiment. The welfare of animals must supercede results and conclusions of the experiment. On the other hand, when an animal is not in good health, the intake of food and water is dramatically reduced, resulting in dehydration and multiorganic failure. As a consequence, experiments where samples are obtained only until the animal's death can give distorted results. The end point of an experiment must be defined before the experiment is carried out and should be evaluated through biochemical parameters, behavior of the animal, the model of the illness, and the treatment. An investigator faces the premature end point of an experiment when the animal has alterations in its behavior for reasons not related to the experiment that modify the expected results, when there is unnecessary suffering and data will not benefit the project, and when the decay in health of the animal causes invalid results. The correct decision about the end point of one experiment suggests that there be permanent monitoring of the animal throughout the experiment to establish knowledge of the behavior and suffering of the animal. The sensible choice of the time for the end point of an experiment, instead of reducing the data of the experiment, will produce better results for supporting the hypothesis or enunciate new ones.

ETHICAL PRINCIPLES OF RESEARCH WITH ANIMALS

Basic ethical principles concerning human health state the importance of knowledge in biology and medicine. As a consequence, experiments involving animal models are necessary. However, the use of animals must imply respect and be humane. Therefore, those who carry out experiments with animals must know that animals also have senses, memory, and are susceptible to pain and suffering.

The researcher is responsible for his actions in the context of a research project; therefore, experiments carried out on animals must be done by qualified researchers or controlled by them. The conditions in which the animals are maintained throughout the experiment must be defined and controlled by a veterinarian or by a competent scientist. The experiment must involve the species of animals that can better adapt to the experiment, and the sensorial and psychological properties of the species involved is also an important factor to be considered before choosing the animals.[7] The investigator must care about the experimental condition of the animal and give necessary help in order to avoid physical and psychological suffering. Furthermore, projects that involve animals must contribute to the knowledge of human health or the well-being of animals or human beings.

From the concepts stated above emerge the principle that working with animals is not a right, but a privilege. Researchers who are involved in the experiments, whether they are assistants or are in charge of the project, must incorporate this principle into their thinking. Because animals lack autonomy and have no choice about participation in an experiment, the researcher must not abuse this privilege.

Although it is possible that animals do not suffer pain in the same way as humans do, there is no reason to suppose that animals do not feel pain and suffer as a result of it.[8]

Abnormal behavior, movements, and postures are signals of pain in experimental animals. Other signs, such as aggressiveness, salivation, unusual sounds, facial expression, etc., are also indicators of pain and suffering. Therefore, the person who deals with experimental animals must be aware of the animal's normal behavior, and also have the ability to detect the minor signs of stress, pain, and suffering.

In summary, the researcher must consider that the well-being of animals is as important as the results of the experiment, and he has an obligation to reduce all the possibilities of suffering pain and stress in the experimental animal.

Animal Well-Being

Well-being is a term based on the human perspective, virtues, and ethical values. However, the well-being of animals involves the absence of pain and stress.[9] The animal needs an appropriate environment for its normal behavior, which can be affected by the animal's senses and perception. In summary, animal well-being is an internal state involving quality of life that is affected by the responses to internal and external stimuli, which may or may not be aversive. It is necessary to establish rules for the care and breeding of animals that will cause the least stress on them. These rules must include all aspects of a normal life for the animals, such as nutrition, housing, feeding, treatment and prevention of illnesses, anesthesia, analgesia, and, when necessary, euthanasia. For example, the stated purpose of the *Guide for the Care and Use of Laboratory Animals* "is to assist institutions in caring for and using animals in ways judged to be scientifically, technically, and humanely appropriate."

The researcher who deals with experimental animals must be knowledgeable about conditions in the area where the animal is housed and where the experiments are carried out. Experimental conditions must be carefully controlled in order to obtain standardized responses. In this way, a smaller number of animals would be involved in the experiment. In addition, the results would be comparable with those from other laboratories around the world. The environmental conditions that must be controlled include:

1. Climate: temperature, humidity, ventilation
2. Physical–chemical: light, sound, presence of contaminants, composition of air, light–dark cycle
3. Rooms: shape, size, number of animals per cage
4. Nutrition
5. Microorganisms and parasites
6. Transport

In regard to the transport of animals, there must be minimum stress on the animals and the travel must not have an impact on their well-being. In addition, they must be secure and comfortable. The stressors on the animals as they travel can be physical (changes in temperature and humidity, sounds, etc.), physiological (access to water and food), and psychological (exposure to new individuals or environments). The effects of stressors are acute and can remain for several days. Acclimation to the new container where the animal will be housed can contribute to the decrease in stress.[10]

In addition, in order to establish universal principles in the practices of ethical care of experimental animals, categories of discomfort to the animal during experimental conditions have been established:[11]

1. Minor discomfort: collection of blood samples, collection of urine in metabolic cages, treatment with drugs in the drinking water, housing in cages in order to observe normal behavior, administration of substances, experimentation with anesthesia and vaccines.
2. Moderate discomfort: frequent sampling of blood, catheterization and intubation, recuperation from general anesthesia as well as immunization with complete adjuvants, cannulation, and recovery from general anesthesia.
3. Severe discomfort: extraction of ascitic fluids, obtaining large volumes of blood without anesthesia, induction of genetic defects, starvation, periods without *ad libitum* access to water, perturbation during periods of sleep, infections, fractures, diabetes, pancreatitis, and renal failure.

A scale of invasiveness of experimental procedures has also been established. It allows the researcher and the ethical committee to evaluate the necessity of special training before the experiment is carried out, to establish standardized operational procedures, to choose from alternative techniques and procedures, and to accompany and supervise the experiments. All of these topics must be evaluated and approved before the experiments can be carried out.

Category	Procedure
A	Experiments with invertebrates, cells or isolated tissues
B	Experiments that cause no stress or minimal discomfort
C	Experiments that cause minimal stress or short duration pain
D	Experiments causing moderate to severe stress or discomfort
E	Procedures involving severe pain in conscious or nonanesthetized animals

This ranking is not only limited to surgical procedures, but it can also include other situations, such as noxious stimuli or agents whose effects are unknown, exposure to drugs or chemicals, behavior studies, nutritional experiments, etc.[10]

CONCLUSION

Although biomedical science research not involving animals is almost impossible, the researcher has a moral obligation to respect the life of the research animal. Apart from the specific objectives of the project, the researcher must avoid unnecessary pain and create the best conditions for housing, sampling, and euthanasia, if necessary. The international harmonization of biological assays is a major effort that is necessary in the research world and animal well being has to be a central issue in current times.

REFERENCES

1. Guide for the care and use of laboratory animals. NIH Publication No. 86-23. Revised 1985.
2. International Council for Laboratory Animal Science (ICLAS), http://www.iclas.org/
3. Federation of European Laboratory Animal Science Associations (FELASA), http://www.felasa.eu/
4. American Association for Laboratory Animal Science (AALAS), http://www.aalas.org/
5. The Canadian Council on Animal Care (CCAC), http://www.ccac.ca/
6. Russell, W.M.S. and R.L. Burch. 1959. *The principles of humane experimental technique*, London: Methuen, p. 238.
7. Repetto, M. 1997. *Toxicología fundamental*. Edition 3. España: Ediciones Díaz de Santos.
8. Cardozo de Martinez, C.A., A. Mrad de Osorio, C. Martínez, E. Rodríguez Yunta, and F. Lolas Stepke. 2007. *El animal como sujeto de experimental. Aspectos técnicos y éticos*. Centro Interdisciplinario de Estudios en Bioética (CIEB). Universidad de Chile, Santiago.
9. Clark, J.D., D.R. Rager, and J.P. Calpin. 1997. Animal well being. I-General Considerations. *Laboratory Animal Science*, 47: 564–569.
10. *Guidelines on procurement of animals used in science*. The Canadian Council on Animal Care (CCAC), http://www.ccac.ca/
11. Mrad, A. 2006. Ética en la investigación con modelos animales experimentales. *Rev. Colombiana de Bioética*, 1(1): 163–83.

2 Control of Stress and Distress

María F. Landoni

INTRODUCTION

There is no doubt that scientific research has improved the quality of life and welfare (reducing suffering and increasing life expectancy) in animals and human beings.

However, the use of animals for research has been a source of controversy for years. In the past few years, the application of Russell and Burch's 3R principles (replace, reduce and refine) (see Chapter 1) has improved the quality of the animal use. However, to imagine scientific research in biology without the use of experimental animals is, at least at the present time, a utopia.

The design of any experimental study in which animals are included should be developed with the aid of a responsible researcher. He must consider, not only his beliefs, but also the ethical consequences of his actions. It should be noted that in animal experimentation, the objective should never justify the means.

Among the important variables that should be evaluated during the experimental design, the most important is the "experimental unit," which in any *in vivo* study is the experimental animal.

Responsible use of laboratory animals is not only a reflection of the ethical background of the researcher, but also his academic and scientific knowledge. Therefore, irresponsible use of laboratory animals has not only ethical consequences, but also serious scientific consequences because it can lead to erroneous results.

There are circumstances during which even a nontrained observer can recognize animal discomfort. However, there are situations in which an untrained person would not recognize discomfort in the research animal, such as stress, distress, and pain.

STRESS

Stress can be defined as the effect produced by external events or internal factors (referred to as stressors), which induce an alteration in the animal's homeostasis. Stressors can be classified according to the cause of the stress:(1) *Physiologic*: pain as a consequence of an injury, surgery or disease, starvation, and dehydration; (2) *Psychological*: fear, anxiety, boredom, loneliness, and separation; and (3) *Environmental*: restraint, noise, odors, habitat, ecology, presence of other species, chemicals, and pheromones. In many cases, the response of an animal to a stressor is adaptive, and when the stressor is eliminated, the animal returns to a state of comfort.

Responses to stressors often involve changes in physiologic function (biochemical, endocrinologic, or autonomic), psychological state, and behavior. Stress is not always harmful to an animal. In some cases, environmental alterations that induce stress also initiate responses that have potential beneficial effects. A minimal degree of stress is necessary for well-being. However, the differentiation between adaptive and nonadaptive stress must be based on professional evaluation and judgment.

The old definitions of stress have emphasized physiologic characteristics, especially those related to neuroendocrine systems. However, in the past few years, it has been demonstrated that stress is not a discrete, well-defined physiologic state.

However, it is generally agreed that some environmental conditions can induce pronounced or persistent stress in an organism and lead to alterations in neuroendocrine activities. The neuroendocrine changes, in some instances, can be severe enough to place the organism in a state of vulnerability to dysfunction or disease, although its behavior might not differ markedly from that typical of its species and should not be considered maladaptive.

Distress

Distress is defined as an aversive state in which an animal is unable to adapt completely to stressors and the resulting stress, and shows maladaptive behaviors. Distress can be the consequence of experimental or environmental stimuli and is identified by behavioral changes, such as abnormal feeding, absence or diminution of postprandial grooming, inefficient reproduction, or inappropriate social interaction with the other animals or handlers (e.g., aggression, passivity, or withdrawal). Distress can also induce pathologic changes that are not directly evident in behavior, such as gastric and intestinal lesions (ulcers), hypertension, and immunosuppression. Maladaptive responses, physiologically designed for reducing distress of animals, can be reinforced, becoming permanent components of the animal's behavior. These compulsive behaviors are dangerous and seriously threaten its well-being. Generally, any behavior that relieves the intensity of distress is likely to become habitual, although its long-term effects will threaten animal well-being. Examples of such behaviors are coprophagy, hair-pulling, self-biting, and repetitive stereotyped movements.

Distress in laboratory animals is unnecessary and unwarranted; therefore, it should be avoided. It is very difficult to precisely identify and measure distress in laboratory animals. Animal users must be trained in the prevention and alleviation of distress. The possibility of distress should be foreseen before laboratory animals are used experimentally; in other words, during the design of the experiment. During the experimental design, stressors that could lead to distress should be identified as well as recognizing the changes in normal behaviors.

Recognition of abnormal behaviors requires that species-typical behaviors associated with well-being be understood and that the normal behavior and appearance of the animals that are being used should be known. Distress can be subtle and can influence experimental outcomes.

Nominal stress is usually a cause for alarm only if an animal is unable to adapt properly to it. When that occurs and distress results, the researcher should identify the underlying cause and begin treatment. Pain-induced stress should then be alleviated by removal of the cause of the pain or through administration of analgesics (see Chapter 3), but nonpain-induced stress is seldom amenable to pharmacological treatment alone.

PHARMACOLOGICAL CONTROL OF STRESS AND DISTRESS

The tranquilizers and sedatives used in animals include drugs in three groups:

1. Major tranquilizers (antipsychotic and neuroleptics):
   a. Phenothiazines
   b. Butyrophenones
2. Minor tranquilizers (antianxiety sedative)
   a. Benzodiazepines
3. 2-Adrenergic agonists

Major Tranquilizers (Antipsychotic and Neuroleptics)

Phenothiazines: Promazine and Acetylpromazine

Phenothiazines are neuroleptic agents. They block postsynaptic dopamine receptors in the central nervous system (CNS) and may also inhibit the release of, and increase the turnover rate of, dopamine. These compounds depress portions of the reticular activating system that assists in the control of body temperature, basal metabolic rate, emesis, vasomotor tone, hormonal balance, and alertness. Phenothiazines also have varying degrees of anticholinergic, antihistaminic, and alpha-adrenergic blocking effects. Phenothiazines can decrease respiratory rates, with little or no effect on blood gas, pH, or oxyhemoglobin saturation.

These drugs have excellent sedative properties as well as antiemetic and antiarrhythmogenic effects. They have no analgesic activity, but when administered with other anesthetics can potentiate their effect. Table 2.1 shows the doses of phenothiazines in various species.

Butyrophenones: Azaperone, Droperidol

The butyrophenones cause tranquilization and sedation (sedation may be lighter than with phenothiazines), antiemetic activity, reduced motor activity, and inhibition of the CNS catecholamines. They have minimal effects on respiration and may inhibit some of the respiratory depressive effects induced by general anesthetics. These compounds can reduce blood pressure as a reflection of their alpha 1 inhibitory activity. However, antagonism of alpha-1-adrenergic receptors is lower than that reported for phenothiazines.

Butyrophenones are mainly used in swine and also as neuroleptics in horses. In swine, they are indicated for the control of aggressiveness, especially when mixing or regrouping weanling or feeder pigs up to 36.4 kg.

As well as phenothiazines, butyrophenones have no analgesic activity, but when administered with other anesthetics can potentiate their effect. Table 2.2 shows the doses of butyrophenones in various species.

TABLE 2.1
Doses of Phenothiazines in Various Species

Species	Drug	Dose mg/kg (administration route)
Rat	Promazine	0.5 – 1.0 (i.m.)
	Acetylpromazine	1 (i.m.)
Mouse	Promazine	5 (i.p.)
	Acetylpromazine	2 – 5 (i.p.)
Guinea pig	Promazine	0.5 – 1 (i.p.)
Hamster	Promazine	0.5 – 1 (i.p.)
Rabbit	Promazine	1 – 2 (i.m.)
	Acetylpromazine	1 (i.m.)
Cat	Promazine	2.2 – 4.4 (i.v., i.m.)
	Acetylpromazine	0.03 – 0.05 (i.v., i.m.)
Dog	Promazine	2.2 – 4.4 (i.m.)
	Acetylpromazine	0.03 – 0.05 (i.m.)
Cattle	Promazine	0.4 – 1.1 (i.v., i.m.)
	Acetylpromazine	0.1 (i.m.)
Sheep/goat	Promazine	0.44 – 1.1 (i.v., i.m.)
	Acetylpromazine	0.04 – 0.06 (i.v., i.m.)
Pig	Acetylpromazine	1.1 – 2.2 (i.m.)
Primate	Acetylpromazine	0.2 (i.m.)

Note: i.m. = intramuscular, i.p. = intraperitoneal, i.v. = intravascular, s.c. = subcutaneous, p.o. = oral, N/A = not available.

TABLE 2.2
Doses of Butyrophenones in Various Species

Species	Drug	Dose mg/kg (administration route)
Rat	Azaperone	3 (i.p.)
	Droperidol	0.5 (s.c.)
Mouse	Azaperone	3.5 (i.p.)
	Droperidol	6 (i.m.)
Guinea pig	Droperidol	0.1 (i.m.) (+ fentanyl)
Hamster	Azaperone	2 (i.m.)
	Droperidol	5 (i.p.)
Rabbit	Azaperone	2 (i.m.)
	Droperidol	0.44 (i.m.) (+ fentanyl)
Cat	Droperidol	0.1 (s.c.) (+ fentanyl)
Dog	Droperidol	0.1 (s.c.) (+ fentanyl)
Cattle	Azaperone	N/A
	Droperidol	N/A
Sheep/goat	Azaperone	N/A
	Droperidol	N/A
Pig	Azaperone	5 (i.m.)
	Droperidol	2.2 (i.m.)
Primate	Droperidol	N/A

Note: i.m. = intramuscular, i.p. = intraperitoneal, i.v. = intravascular, s.c. = subcutaneous, p.o. = oral, N/A = not available.

Minor Tranquilizers (Antianxiety Sedative)

Benzodiazepines: Diazepam, Midazolam

Benzodiazepines are depressants of the CNS at the subcortical levels (primarily limbic, thalamic, and hypothalamic) leading to anxiolytic, sedative, skeletal muscle relaxant, and anticonvulsant effects. These compounds act as modulators or facilitators of gamma amino butyric acid (GABA) activity, by binding a specific receptor located in the GABA-dependent chloride channel.

Benzodiazepines are used clinically for their anxiolytic, muscle relaxant, hypnotic, appetite stimulant, and anticonvulsant activities. These compounds have a wide therapeutic window; however, in cats, they can induce changes in behavior (irritability, depression, and aberrant demeanor), especially after administration of diazepam.

Diazepam has a long half-life in most species and is indicated for reducing hyperexcitability and anxiety. Midazolam has a shorter half-life and, therefore, is indicated as a premedicant for general anesthesia or in circumstances where rapid and short-lasting effect is required. Table 2.3 shows the doses of benzodiazepines in various species.

2-Adrenergic Agonists

Xylazine, Medetomidine

2-Adrenergic agonists are potent sedative/analgesics with muscle relaxant properties. Compared to opioids, alpha-2-agonists do not cause CNS excitation in any species. These compounds cause muscle relaxation through central mediated pathways. They also depress thermoregulatory

TABLE 2.3
Doses of Benzodiazepines in Various Species

Species	Drug	Dose mg/kg (administration route)
Rat	Diazepam	3 – 5 (i.m.)
	Midazolam	1 – 2 (i.m.)
Mouse	Diazepam	3.5 (i.m.)
	Midazolam	1 – 2 (i.m.)
Guinea pig	Diazepam	3 – 5 (i.m.)
	Midazolam	1 – 2 (i.m.)
Hamster	Diazepam	3 – 5 (i.m.)
	Midazolam	1 – 2 (i.m.)
Rabbit	Diazepam	5 – 10 (i.m.)
	Midazolam	1 – 2 (i.m.)
Cat	Diazepam	0.1 – 0.25 (s.c., i.m.)
	Midazolam	0.2 – 0.4 (i.m.)
Dog	Diazepam	0.1 – 0.25 (s.c., i.m.)
	Midazolam	0.2 – 0.4 (i.m.)
Cattle	Diazepam	0.4 (i.m.)
	Midazolam	N/A
Sheep/goat	Diazepam	0.55 – 1 (i.m.)
	Midazolam	N/A
Pig	Diazepam	0.5 – 10 (i.m.)
	Midazolam	1 (i.v., i.m.)
Primate	Diazepam	0.5 – (p.o.) 0.25 – 0.5 (i.m.)
	Midazolam	3 (p.o.) 0.1 – 0.5 (i.m.)

Note: i.m. = intramuscular, i.p. = intraperitoneal, i.v. = intravascular, s.c. = subcutaneous, p.o. = oral, N/A = not available.

mechanisms, and either hypothermia or hyperthermia is a possibility, which will depend on ambient temperature.

Effects on the cardiovascular system include an initial increase in peripheral resistance, with the consequent increase in blood pressure followed by a longer period of lowered blood pressure. A bradycardia effect can be observed with, in some animals, development of second-degree heart blockade or other arrhythmias. Xylazine has been reported to enhance the arrythmiogenic effects of adrenaline. Medetomidine is more specific than xylazine for alpha-2-receptors versus alpha-1-receptors; therefore, effects on blood pressure and heart are lighter. Effects of alpha-2-agonists in the respiratory system are minimal and clinically insignificant.

Alpha-2-agonists are indicated for reducing hyperexcitability, producing sedation with a short period of analgesia, and as a preanesthetic for local and general anesthesia. Table 2.4 shows the doses of alpha-2-agonists in various species.

NONPHARMACOLOGICAL CONTROL OF STRESS AND DISTRESS

Nonpharmacological approaches to preventing, minimizing, and alleviating (nonpain induced) stress and distress in animals include good husbandry and management practices, socialization and handling, and environmental enrichment.

TABLE 2.4
Doses of alpha-2-Agonists in Various Species

Species	Drug	Dose mg/kg (administration route)
Rat	Xylazine	1 – 3 (i.m.)
	Medetomidine	0.25 – 0.5 (i.m.)
Mouse	Xylazine	13 (i.p.)
	Medetomidine	0.25 – 0.5 (i.m.)
Guinea pig	Xylazine	8 – 10 (i.m.)
	Medetomidine	0.5 (i.m.)
Hamster	Xylazine	8 – 10 (i.m.)
	Medetomidine	0.5 (i.m.)
Rabbit	Xylazine	5 (i.m.)
	Medetomidine	0.25 – 0.5 (i.m.)
Cat	Xylazine	1.1 – 2.2 (s.c., i.m.)
	Medetomidine	0.04 – 0.08 (i.m.)
Dog	Xylazine	1.1 – 2.2 (s.c., i.m.)
	Medetomidine	0.01 – 0.04 (i.m.)
Cattle	Xylazine	0.05 – 0.15 (i.m.)
	Medetomidine	N/A
Sheep/goat	Xylazine	0.05 – 0.1 (i.m.)
	Medetomidine	N/A
Pig	Xylazine	1 – 2 (i.m.)
	Medetomidine	N/A
Primate	Xylazine	0.25 – 2 (i.m.)
	Medetomidine	0.02 – 0.06 (i.m.)

Note: i.m. = intramuscular, i.p. = intraperitoneal, i.v. = intravascular, s.c. = subcutaneous, p.o. = oral, N/A = not available.

Husbandry and Management Practices

Husbandry and management practices in animal care and housing are important sources of stressors, such as anxiety, boredom, fear, and loneliness. These stressors are potential causes of distress and development of maladaptive behaviors. To avoid or minimize distress is fundamental to knowing and understanding the social and physical needs of the different laboratory animals. These requirements are accessible to the public on the Web site of FELASA (www.felasa.eu).

With regard to housing, potential environmental stressors that can lead to stress and distress (e.g., levels of ambient light, noise, vibrations, temperature, and disturbances from operation of facilities) should be kept to a minimum. It is true that no environment is free of stressors; however, it is important to respect the international guidelines for care and use of laboratory animals. It is generally preferable to house animals that are social by nature, such as rats, mice, dogs, and primates, in groups (unless there are scientific or welfare reasons not to do so). Social housing among compatible individuals is neither stressful nor harmful. Furthermore, evidence indicates that housing naturally sociable animals in solitary conditions can result in distress and harm.

Many techniques that minimize stress in husbandry, such as combining husbandry handling with habituation and handling for research purposes, acclimation to new environments, positive reinforcements, operant conditioning, and well-trained staff, can be helpful tools for the overall reduction of stress and distress. Environmental enrichment can improve animal welfare, reduce stress, and improve the quality of data obtained from the animals in situations where it does not compromise the anticipated research outcomes by introducing uncontrolled or unanticipated variables.

BIBLIOGRAPHY

AVTRW (Association of Veterinary Teachers and Research Workers). 1986. Guidelines for the recognition and assessment of pain in animals. *The Veterinary Record* 118: 334–338.

Flecknell, P.A. 1984. The relief of pain in laboratory animals. *Laboratory Animals* 18: 147–160.

Flecknell, P.A., and J.H. Liles.1992. Evaluation of locomotor activity and food and water consumption as a method of assessing postoperative pain in rodents. In *Animal Pain,* Eds. C.E. Shortland and A. Van Poznak. New York: Churchill Livingstone, pp. 482–488.

LASA (Laboratory Animal Science Association). 1990. The assessment and control of the severity of scientific procedures on laboratory animals. *Laboratory Animals* 24: 97–130.

Liles, J.H., and P.A. Flecknell. 1992. The use of non-steroidal anti-inflammatory drugs for the relief of pain in laboratory rodents and rabbits. *Laboratory Animals* 26: 241–255.

Melzack, R., and S.G. Dennis. 1980. Phylogenetic evolution of pain-expression in animals. In *Pain and Society*, Eds. H.W. Kosterlitz and L.Y. Terenius. Wheinheim, Basel: Verlag Chemie, pp. 13–26.

Morton, D.B., and P.H.M. Griffiths. 1985. Guidelines on the recognition of pain, distress and discomfort in experimental animals and an hypothesis for assessment. *Veterinary Record* 116: 431–436.

Wright, E.M., K.L. Marcella, and J.F. Woodson. 1985. Animal pain: Evaluation and control. *Laboratory Animals* 14: 20–30.

3 Management of Pain in Laboratory Animals

María F. Landoni

INTRODUCTION

Pain in laboratory animals is a major animal welfare problem that must be addressed. In order to provide effective analgesia, it is essential to have a good knowledge and understanding of animal pain. It is fundamental to know when pain might occur, how long it might last, and how well it will respond to therapy. It is also essential to consider the advantages and disadvantages of the various methods of managing pain, and which is the best in different situations.

The golden rule for proper pain management is to be capable of recognizing the presence of pain and to assess its severity. The capacity of animals to experience sensations, such as pain, in a similar way as man has been, and still is, extensively debated. The main reason for the continued debate is the difficulty of directly investigating such emotional states. Scientific support for the belief that animals can suffer often consists simply of drawing parallels in animal and human neuroanatomy, and making the assertion that animals "are given the benefit of the doubt." The prefrontal cortex has been reported as associated with pain perception or at least with the emotional components of pain. In fact, it has been demonstrated that humans undergoing a prefrontal lobotomy (used in the past as a treatment for some psychiatric disorders) still responded to painful stimuli by reflex movements, but expressed no concern about the pain; it was no longer considered unpleasant. Most animal species have relatively small areas of prefrontal cortex, and this has led to the suggestion that pain in animals is comparable to that experienced by lobotomized humans. This assumes that the actual size of the prefrontal cortex will determine the capacity for pain perception, but it may be that other areas of the brain carry out a similar role in other species. What is clear, however, is that animals do not behave in simple, reflexive ways—in circumstances that would cause pain in man.

The development of proper assessment techniques is essential for understanding and appreciation of animal pain. If pain cannot be recognized and assessed it cannot be effectively managed. One of the most common mistakes when trying to recognize animal pain is to expect animals to behave as human beings. Animals in pain will behave in different ways depending upon the site, severity, and type of pain. Also, they will behave in a species-specific way. Some species, especially those that may expect support from others, may show very obvious pain-related behavior. In other species, expressing such overt behavior would simply alert predators that they were less fit and, hence, easy prey. If overt pain-related behavior is expressed, then the animal may mask this behavior when it is aware it is being observed. Animals may also change their responses when in a familiar, secure environment, and express less pain-related behavior than when in an unfamiliar environment, e.g., when removed from their cage for examination.

A proper pain assessment requires:

1. Adequate knowledge of the species-specific behaviors of the animal being assessed
2. Adequate knowledge to be able to compare the individual animal's behavior before and after the onset of pain (e.g., pre- and postoperatively)
3. Proper clinical examination of the affected area

At present, in veterinary clinical practice, analgesics are widely used to control pain in two groups of animals that have undergone surgery or have suffered traumatic injuries, and those with acute or chronic arthritis. In research animal facilities, alleviation of postoperative pain probably represents the greatest area of analgesic use. In this context, there are some myths related to the use of analgesics that have been addressed by Paul Flecknell. They include:

1. Alleviation of postoperative pain will result in the animal injuring itself.
2. Analgesic drugs have undesirable side effects, such as respiratory depression.
3. We don't know the appropriate dose rates and dosage regimens.
4. Pain-relieving drugs might adversely affect the results of an experiment.

Pain can be relieved by the so-called analgesics, which can be broadly divided into two groups: the opioids or narcotic analgesics and the nonsteroidal antiinflammatory drugs (NSAID), such as aspirin. Local anesthetics can also be used to provide postoperative pain relief by blocking all sensation in the affected area.

Opioids, especially potent μ agonists (morphine, oxymorphone, fentanyl, petidine), have a relatively short duration of action; therefore, maintenance of effective analgesia may require repeated administration every one to three hours, depending on the species. Partial μ-agonists, such as buprenorphine, have a longer duration of action and also can be administered orally (does not suffer first pass effect). NSAIDs tend to have long duration of action, and the newer agents (e.g., carprofen, ketoprofen, and meloxicam) are effective in controlling moderate postsurgical pain in many circumstances. Doses and administration intervals for the analgesics most commonly used in laboratory animals are presented in Table 3.1 and Table 3.2.

The most effective analgesia is observed when analgesics are administered preoperatively. In the past few years, multimodal pain therapy has been used. Multimodal pain therapy consists of administering several different classes of analgesics, each acting on different parts of the pain system (e.g., by combining the use of opioids (acting centrally by limiting input of nociceptive information into the CNS) with NSAIDs (acting peripherally, by decreasing inflammation during and after surgery, and thus limiting the nociceptive information entering the CNS).

In summary, pain in laboratory animals must be alleviated and in doing that, it is essential to recognize its presence. It also is important to note that alleviation of postoperative pain will be ineffective if the applied surgical techniques are not used properly, as well as, if during recovery from anesthesia, animals remain in uncomfortable places (e.g., wet, cold, noisy) because such environments are likely to cause distress.

TABLE 3.1
Doses and Administration Intervals for the Commonly Used Opioid Analgesic in Laboratory Animals

	Buprenorphine	Butorphanol	Morphine	Nalbufine	Pethidine (Meperidine)
Rat	0.01–0.05 mg/kg s.c., i..v. q/8–12 h	1–2.0 mg/kg i.m. s.c. q/2–4 h	2–5 mg/kg i.m., s.c. q/4 h	1–2 mg/kg i.m. q/4 h	10–20 mg/kg i.m., s.c. q/2–3h
Hamster	0.01–0.03 mg/kg i.m., s.c., i.v. q/6–12h	0.01–0.05 mg/kg i.m.,s.c., i.v. q/6–12 h	0.5–2 mg/kg i.m., s.c. q/4–6 h	N/A	5–10 mg/kg i.m. q/2–4 h
Guinea pig	0.05 mg/kg s.c. q/8–12 h	0.4 mg/kg i.m. q/4 h	2–5 mg/kg i.m., s.c. q/4 h	1–2 mg/kg i.m. q/4 h	10 mg/kg i.m. q/2–4 h
Mouse	0.05–0.1 mg/kg s.c. q/8–12 h	1–2.0 mg/kg i.m. s.c. q/4 h	2–5 mg/kg i.m., s.c. q/4 h	2–4 mg/kg i.m. q/4 h	10–20 mg/kg s.c., i.m. q/2–3h
Rabbit	0.1–0.25 mg/kg p.o. q/8–12 h	0.1–0.5 mg/kg i.m., s.c., i.v. q/4 h	2–5 mg/kg i.m., s.c. q/4 h	1–2 mg/kg i.m., i.v. q/4 h	5–10 mg/kg i.m., s.c. q/2–3h
Primates	0.005–0.01 mg/kg i.m., s.c., i.v. q/6–12 h	0.01 mg/kg i.v.	0.1–2 mg/kg i.m., s.c. , i.v. q/4–6 h	N/A	N/A
Dog	0.005–0.02 mg/kg i.m., s.c., i.v. q/6–12 h	N/A	0.1–1.0 mg/kg i.m., s.c., i.v. q/4–6 h	0.3–0.5 mg/kg i.m., s.c. q/3–4 h	3.5–10 mg/kg i.m. o 10–15 mg/ kg s.c. q/2.5–3.5 h
Cat	0.005–0.02 mg/kg i.m., s.c., i.v. q/6–12 h	N/A	0.1–0.2 mg/kg i.m., s.c., i.v. q/6–8 h	0.3–0.5 mg/kg i.m., s.c. , i.v. q/3 h	3.5–10 mg/kg i.m. o 10–15 mg/ kg s.c., q/2–3 h
Pig	0.005–0.05 mg/kg i.v., i.m. q/6–12 h	0.2–0.6 mg/kg i.m., s.c., i.v. 2-q/4 h	0.2–1.0 mg/kg i.m. q/4 h	N/A	2 mg/kg i.m., i.v. q/2–4 h
Sheep/goat	0.005–0.01 mg/kg i.v., i.m. q/4 h	0.2–0.8 mg/kg i.m., s.c. q/2–4 h	0.2–0.5 mg/kg i.m. q/2 h	N/A	2 mg/kg i.m., i.v. q/2 h

Note: i.m. = intramuscular, i.p. = intraperitoneal, i.v. = intravascular, s.c. = subcutaneous, p.o. = oral, N/A = not available.

TABLE 3. 2
Doses and Administration Intervals for the Commonly Used NSAIDs in Laboratory Animals

	Aspirin	Carprofen	Diclofenac	Flunixin	Ketoprofen
Rat	100 mg/kg p.o. Single dose	5 mg/kg s.c., p.o. q/24 h	10.0 mg/kg p.o. q/24 h 50 mg/kg i.m. q/8 h	2.5 mg/kg s.c. q/12—24 h	5 mg/kg s.c. u p.o. q/24 h
Hamster	200 mg/kg p.o. Single dose	N/A	N/A	0.5-2 mg/kg s.c., q/12–24 h	2 mg/kg s.c. q/24 h (no more than 3 days)
Guinea pig	80–90 mg/kg p.o. Single dose	N/A	2.0 mg/kg p.o. q/24 h	N/A	N/A
Mouse	120 mg/kg p.o. Single dose	5 mg/kg s.c. p.o. q/24 h	8.0 mg/kg p.o. q/24 h	2.5 mg/kg s.c. q/12-24 h	N/A
Rabbit	100 mg/kg p.o. Single dose	4 mg/kg s.c. q/24 h o 1.5 mg/kg p.o. q/12 h	N/A	1.0 mg/kg s.c. q/12–24 h	3 mg/kg s.c. q/24 h
Primates	20 mg/kg p.o. q/6-8 h	2–4 mg/kg s.c., i.v., single dose	2–4 mg/kg s.c., i.v. q/24 h 1 mg/kg S.C., IV single dose	0.2 mg/kg s.c., single dose	2 mg/kg s.c. q/24 h (no more than 3 days)
Cat	10–25 mg/kg p.o. q/24–48 h	2 mg/kg p.o., for 4 days, afterwards every other day	1 mg/kg p.o. single dose 1 mg/kg s.c., i.v. single dose	0.3 mg/kg p.o.	1 mg/kg p.o. q/24 h (no more than 5 days)
Dog	10–25 mg/kg p.o. q/24h	4 mg/kg s.c. o i.v., single dose	1 mg/kg p.o., q/24 h (no more than 3 days)	0.2 mg/kg s.c. single dose	2 mg/kg s.c., q/24 h (no more than 3 days)
Pig	50–100 mg/kg p.o. q/6-12 h	4 mg/kg p.o.	1–2 mg/kg s.c., i.v. q/24 h	0.1–0.2 mg/kg p.o. q/24 h	1 mg/kg p.o. q/24 h (no more than 5 days)
Sheep/goat	N/A	2–4 mg/kg s.c., i.v. q/24 h	2 mg/kg s.c., i.v. q/24 h	N/A	N/A

Note: i.m. = intramuscular, i.p. = intraperitoneal, i.v. = intravascular, s.c. = subcutaneous, p.o. = oral, N/A = not available.

BIBLIOGRAPHY

AVTRW (Association of Veterinary Teachers and Research Workers). 1986. Guidelines for the recognition and assessment of pain in animals. *The Veterinary Record* 118: 334–338.

Flecknell, P.A. 1984. The relief of pain in laboratory animals. *Laboratory Animals* 18: 147–160.

Flecknell, P.A., and J.H. Liles. 1992. Evaluation of locomotor activity and food and water consumption as a method of assessing postoperative pain in rodents. In *Animal Pain*, Eds. C.E. Shortland and A. Van Poznak. New York: Churchill Livingstone, pp. 482–488.

LASA (Laboratory Animal Science Association). 1990. The assessment and control of the severity of scientific procedures on laboratory animals. *Laboratory Animals* 24: 97–130.

Liles, J.H., and P.A. Flecknell. 1992. The use of non-steroidal anti-inflammatory drugs for the relief of pain in laboratory rodents and rabbits. *Laboratory Animals* 26: 241–255.

Melzack, R., and S.G. Dennis. 1980. Phylogenetic evolution of pain-expression in animals. In *Pain and Society*. Eds. H.W. Kosterlitz and L.Y. Terenius. Wheinheim, Basel: Verlag Chemie, pp. 13–26.

Morton, D.B., and P.H.M. Griffiths. 1985. Guidelines on the recognition of pain, distress and discomfort in experimental animals and an hypothesis for assessment. *Veterinary Record* 116: 431–436.

Wright, E.M., K.L. Marcella, and J.F. Woodson. 1985. Animal pain: Evaluation and control. *Laboratory Animals* 14: 20–30.

4 Anesthesia and Analgesia

Lucas F. de Candia, Alfredo Rigalli, and Verónica E. Di Loreto

GENERAL ANESTHESIA

The main aim of anesthesia in the laboratory animal is to produce the highest degree of analgesia possible in order to prevent the animal from feeling pain. An additional function of the anesthesia is to be able to have control in the manipulation of the animal along with proper muscular relaxation, so that the surgical procedures can be performed according to the techniques applicable in each case.

Good anesthesia is an important aspect in the well-being of the animal and all the investigators should know their ethical and legal responsibilities to avoid unnecessary pain and stress in the animal. It is also relevant as regards the scientific validity of any study that requires animals. On the other side, it plays an important role in the experimental design and results of the biomedical investigation. Therefore, the election of anesthesia and the ways of administration should be carefully considered.[1-3]

TERMINOLOGY

Analgesia: Lack of sensitivity to pain, without causing unconsciousness.
Sedation: Calm state, generally accompanied by drowsiness.
General anesthesia: Temporal controlled unconsciousness, induced through intoxication of the central nervous system.
Time of the maximum effect: Time between the moment of the application of the anesthetic and the moment of its highest effect.
Recovery time: Time between the initial application and the ability of standing without any help.

CHOICE OF THE ANESTHETIC

Any technique can be considered adequate if it allows for absence of pain in the animal, appropriate surgical conditions, and fast recovery. Some aspects of anesthesia include:

- Nature and duration of the surgical procedure: An experiment may require the recovery of the animal or it may not. This can determine the kind of anesthetic used. When the anesthesia period is short (up to 30 minutes) and the surgical procedures are limited, the main factor that influences the choice of the anesthetic is the ease for induction. In this case, it is common to use injectable anesthetics such as thiopental or ketamine. On the other hand, many experimental protocols require excessively long anesthesia periods and the rat is euthanized at the end of the experiment. In these situations, urethane can be used.
- Interference with the results: The anesthetic chosen must cause the least interference possible on the studied systems or on the variables measured in the experiment, in order not to affect the interpretation of the results. For example, if hepatectomy is performed, it is not possible to use an anesthetic that is cleared by the liver. It is important to consider that a

drug may not be the best, but others that are potentially better are not acceptable because of their negative influence on the experiment.

- Standardization of the factors involved: It is important to standardize the various biological factors that may affect the sensitivity or the dose-response: weight, age, sex, strand differences, etc. That is why, when an anesthetic is used for the first time, it is important to see the effect produced on one animal before working with a large number. It is also necessary to consider the other drugs the animal may be receiving as part of the experimental protocol and that may affect the response to anesthetics.

Thus, it is important to evaluate the ways that are available in the laboratory to administrate a certain type of anesthetic as well as the cost of the anesthetics.

Other Factors for Consideration

It is important for the animal to be calm to aid in the induction of the anesthesia. A stressed animal increases the risk of complications and/or death from the effects of anesthetics. Generally, most of the stress of the rat during the administration of anesthesia is due to the improper handling of the animal. Therefore, it is extremely important to be expertly trained and skillful in the handling of the rat in order to avoid complications. If, when the animal is going to be anesthetized, it shows an aggressive behavior, it is recommended to tip the cage and open the door, so that the animal can exit by itself. Once the rat begins to recognize its environment, it should calm down. At this point, it is okay to take the animal and put it on the forearm. It is best to put the animal directly on the bare arm or to use latex gloves. Using gloves made of an excessively rigid material is not recommended. The fact that sometimes the rat tries to bite is generally due to the inexperience or fear of the surgeon.

Rats do not need to be fasting for the induction of the anesthesia, unless the surgery is in the gastrointestinal tract or the experiment requires the absence of food.

Even though some laboratories lack sophisticated equipment, it is important to have some basic tools that ensure adequate ventilation, such as, endotracheal tubes, oxygen source, and oral secretions extractor.

Stages and Levels of Anesthesia

General anesthesia is divided in stages and levels.

Stage 1: Analgesia.
Stage 2: Certain degree of agitation and/or erratic movements can be observed.
Stage 3: Corresponds to the surgical level of anesthesia and is subdivided in levels:
Level 1: Loss of the optical facial winking reflex.
Level 2: Stop of eye movement. The animal presents deep and regular breathing. It is a good level to begin with the surgery.

Stage 4: There is a complete loss of breathing movements, cyanosis, and heartbeats.

Mechanisms of Action of General Anesthetic Drugs

Although general anesthesia is widely used in medical and veterinary practices (as well as in the biomedical studies that use animals), nowadays, the pharmacodynamics of these drugs and, mainly, the relationship between molecular action and cellular mechanisms and the clinically observed effects are not well known.[4]

The membrane receptors that are considered the most important action areas of general anesthetics are members of a ligand-gated, ion channel superfamily, and are widely distributed on the central nervous system (CNS).[5-8] However, some isoforms of receptors depend on the activities of the voltage-dependent ion channels,[7] but it is not known specifically how relevant they are to general anesthesia.

There are two basic methods to perform general anesthesia: injection and inhalation.

Injectable Anesthetics

Injectable anesthetics are preferred in laboratories for several reasons: low cost, simple administration, and no sophisticated equipment is needed to induce anesthesia. Besides, it is not necessary to be extremely concerned about the safety margins, which can affect inhalatory anesthetics.

The disadvantage is that, when it is administrated, the anesthetic remains inside the organism until it is metabolized or eliminated.

As happens with other drugs, the individual variability plays an important role in the bioavailability as well as in the pharmacokinetics. The components of this variability include genotype (species, lineage, strand, etc.), sex, age, body composition, and nutritional status. On the other hand, anesthesiology is not an exact science; therefore, the recommendations and doses given by different sources should be taken as points of reference.

For injectable anesthesia, any route of administration can be used. The intravenous route has an advantage in that it allows a better control of the anesthesia and it is easier to adjust the dose to the individual response of the animal. But its disadvantage is that it may be difficult to administer in rats due to the dimensions of the veins and the great mobility of these animals. The intraperitoneal route, however, has the advantage of being easy to administrate and less painful. It is also possible to use the intramuscular and subcutaneous routes.

There are three different kinds of injectable anesthetics: hypnotic, barbituric, and dissociative.

Hypnotic Anesthetics

Hypnotic anesthetics are CNS-depressant drugs that are not necessarily selective. They produce a dose-dependent response varying from sedation and sleep (hypnosis) to unconsciousness and surgical anesthesia. It can lead to coma and death due to the depression of the respiratory and/or cardiovascular control centers. Examples include urethane (ethylcarbamate), chloral hydrate (trichloro acetaldehyde monohydrate), alpha-chloralose, and tribromoethanol.

Urethane—The ethyl ester of the carbamic acid is easily soluble in water. After administration, it has a wide safety margin and produces long-life narcosis and hypnosis (8 to 10 h) with a minimum of respiratory or cardiovascular depression and maintenance of spinal reflexes. This substance produces stable but superficial anesthesia, which does not allow surgeries in which numerous manipulations produce intense pain in the animal. This can be improved by combining it with the administration of a local anesthetic, such as lidocaine chlorhydrate (see anesthetics combination in this chapter).

It is a medium-term tisular toxic and may produce death. That is what makes it better for surgeries in which the animal is sacrificed at the end of the experiment.

Mechanism of action: There is no universally accepted mechanism to explain the pharmacodynamics of urethane.[9] The action of this drug is speculated to be, in part, the result of the ethanol metabolism that, like barbiturics and benzodiazepines, has an anxiolytic and sedative/hypnotic effect. It can also affect the gamma amino butyric acid (GABA) receptors.[10] However, compared with other anesthetics, urethane has low stimulatory effect on GABAergic transmission pathways.[11]

Advantages: It is administrated in a unique dose. The induction of the anesthetic effect is fast (1 to 2 minutes).

Disadvantages: It is carcinogenic and eye, skin, and mucosa irritation can occur. It must be handled carefully while preparing and administering a solution.

Dose: For i.p. (intraperitoneal) administration, the dose used is 120 mg/100 g of body weight in a solution 120 g/l (see chapter 45).

Experimental results: The percentage of animals surviving after being anesthetized with urethane, in a dose of 120 mg/100 g of body weight is shown in Figure 4.1. In the figure, the percentages according to the different types of surgeries are also shown. As one can see, almost 10% die at the beginning of the surgery; this can be attributed to surgical complications. Then, the percentage varies gradually. It is evident that the death percentage is extremely influenced by the complexity of the surgery. This conclusion comes from comparing the surgeries of vesical catheterization, with vesical catheterization, plus laparotomy.

Barbituric Anesthetics

Barbituric anesthetics are classified as sedatives/hypnotics, which show the dose-dependent capacity to produce sedation or a deep stage of hypnosis. They are depressors of the central nervous system, but they are poor analgesics.

According to the duration of their action, they can be classified as:

1. Short-action barbiturics: pentobarbital
2. Extremely short-action barbiturics: thiopental, methoexital
3. Stable and long-action barbiturics: tiobutabarbital

Pentobarbital and thiopental are used in most cases.

The U.S. Drug Enforcement Administration (DEA) defines these as category II controlled substances (high dependence potential) and a DEA license is needed to purchase them.

Mechanism of action: Barbituric anesthetics, in clinical concentrations, act on the $GABA_A$ receptors.[12] As previously stated, the pharmacological stimulation of these receptors is one of the most relevant points in the anesthesia state because the nervous transmission inhibits several circuits of the CNS.

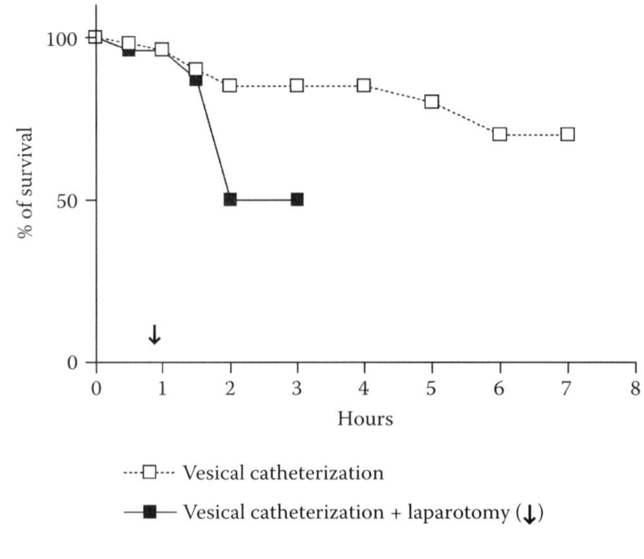

FIGURE 4.1 Survival during surgery.

Pentobarbital

Pentobarbital is an extremely effective barbituric that allows for the recovery of the animal and is generally used in short surgeries. When used for long procedures, it is frequently necessary to use a maintenance dose. Thiopental is an anesthetic similar to pentobarbital.

Dose: Pentobarbital 5 mg/100 g of body weight and thiopental 7 mg/100 g of body weight, i.p. (intraperitoneally).

Dissociative Anesthetics

Ketamine

Utility: It is the most used anesthetic in veterinary practices, probably due to its wide range of safety and compatibility with other drugs. It can be administered intravenously (i.v.), subcutaneously (s.c.), intramuscularly (i.m.), and intraperitoneally (i.p.). It is considered a powerful analgesic that blocks the conduction of the pain impulse toward the thalamic and cortical areas. It has been shown that analgesia is more potent for procedures that involve the muscle skeletal system than for abdominal viscera. In surgical anesthesia levels, the swallowing reflex is maintained, but the optic facial winking reflex is lost. It is used for short duration anesthesia (15 to 20 minutes).

It presents a high solubility in lipids with a fast induction to anesthesia and a return to consciousness following the redistribution to other body tissues. It has a high bioavailability after i.v. or i.m. administration.

In most species it is metabolized through the liver via cytochrome p450. After an i.p. injection of ketamine, 2-week old rats have an anesthesia period longer than adults and, for the same age, greater in females than males.

Mechanism of action: It acts mainly by inhibiting the stimulatory receptor N-methyl-D-aspartate (NMDA), whose ligand is the neurotransmitter glutamate and increases the activity of the receptor of opiods in the encephalic stem[8] that activates the physiological mechanisms of anesthesia. As well as barbiturics, ketamine depresses the cholinergic and nicotinic neural receptors. This inhibition is considered to be one of the factors involved in the induction of the anesthetic state.[5]

Procedure: It is usually combined with tranquilizer and sedative agents to improve the muscular relaxation and analgesic properties as well as to assure a calm recovery. The most frequent combination is with diazepam or xilazine.

Dose: The dose used is 50 mg/kg of body weight combined with 5 mg of diazepam/kg of body weight or 5 mg of xilazine/kg of body weight intraperitoneally. Intramuscularly, the dose is 30 mg/kg of body weight combined with 3 mg of xilazine/kg of body weight. If it is necessary, during surgery, a maintenance dose can be added.

Anesthetic Combinations

Many of the drugs that have been discussed lack one or several of the general anesthesia properties: hypnosis, analgesia, or muscular relaxation. This had led to the improvement of anesthesia quality through the combination of two or more drugs. The combinations used the most in experimental animals are, as mentioned before, ketamine combined with diazepam or xilazine. They improve anesthesia, the muscular relaxation, and sedation along with providing a more lasting anesthesia with a decrease of adverse effects.

Xilazine is an agonist alpha-2 analgesic, a sedative, and it relaxes the skeletal muscle. It is not an anesthetic. It absorbs quickly after i.m. or i.p. administration and it is eliminated relatively fast.

Diazepam is a benzodiazepine used for preanesthesia and anesthesia induction together with another agent (e.g., ketamine), although it is capable of generating anesthesia by itself. It is a sedative, hypnotic, anxiolytic, relaxer of the skeletal muscle, and it has anticonvulsive properties. It is lipophilic with a half-life of 7 h in rats. It is mainly eliminated through extrahepatic functions, and its cardiovascular and respiratory effects are minimal.

Inhalatory Anesthetics

Mechanism of action: Several studies state that all inhalatory anesthetics produce amnesia and inhibit the organism's motor response to nociceptive stimulus. The evaluation of other behaviors and physiological responses showed variable results. The suppression of the motor response is performed by the spinal cord. The hypnosis and amnesia are the consequence of the action of these drugs in the brain. Their mechanism of action is related to the alteration of the ionic channels of the brain cells. There is evidence that they affect the ionic channels by directly joining polypeptide motifs of the receptor. Neuroanatomical differences in distribution of ionic channels and the polymorphism of them are associated to different clinical effects.[13]

Specifically on *in vitro* studies, it was observed that volatile anesthetics like halothane and isofluorane inhibit the glutamaergic and neural cholinergic nicotinic activity and stimulate the GABAergic and glycine channel activity in the CNS.[13-15] In this way, these drugs depress the excitatory nervous network and increase the conductance of the inhibitory ones, leading the brain cells involved to the hyperpolarization of their membranes.

Sulphuric Ether

Utility: Although sulphuric ether has been replaced by other anesthetics it can be useful in some procedures, e.g., euthanasia and heart puncture.

Advantages: It is a good analgesic and muscle relaxer, with a simple administration and low cost.

Disadvantages: The control of the anesthesia is difficult. It irritates the respiratory tract mucosa. It can also stimulate the sympathetic nervous system, which leads to an increase of catecholamines in the rats. Besides, it is extremely important not to use it in closed or unventilated places or around fire because it is very flammable.

Procedure: To perform anesthesia, it is necessary to have an anesthetic container (volume for 3 to 4 l) with a cap (Figure 4.2).

The rat is placed in the container together with a piece of cotton soaked with sulphuric ether (approximately 2 to 3 ml). The anesthetic effects of sulphuric ether appear after 1 to 2 minutes. After this period, it is convenient to leave the animal inside the container for 20 or 30 more seconds. During the procedure, to maintain the anesthesia, it is necessary to put a tube, mask, or funnel with a piece of cotton soaked with ether on the snout of the animal (Figure 4.3). The surgeon must pay attention to the rat's breathing, and move the tube closer to or away from the animal, decreasing or increasing the amount of anesthetic administrated.

Sulphuric ether is an irritant and causes abundant secretions in the superior respiratory tract. As a consequence, it is possible to cause a respiratory failure if there is an excess of anesthetic. If that happens, it will be necessary to perform artificial respiration and extract the mucus. Another option to decrease the signs of respiratory irritation is to use it combined with an anticholinergic agent like atropine.

FIGURE 4.2 Container for rapid anesthesia with sulphuric ether.

Methoxyfluorane

An anesthetic container can be used, but it has the disadvantage of being expensive and nephrotoxic. It is used infrequently nowadays.

Halothane

It is necessary to use a vaporizer. It produces a fast induction to anesthesia (stabilization at 10 minutes). This is why the administration must be carefully controlled. It has a low toxicity due to its molecular stability.

Isofluorane

It is necessary to use a vaporizer. It is an excellent anesthetic to use during investigation, but a lot of knowledge and training in animal monitoring is necessary to avoid an overdose. It produces a fast induction to anesthesia and has a high molecular stability; this is why its toxicity is low. It maintains the cardiovascular function in a better way than halothane and, therefore, it is safer in this aspect.

LOCAL ANESTHESIA

The local anesthetics such as lidocaine chlorhydrate and benzocaine are drugs that block nerve transmission. The most commonly used local anesthetics do present common patterns, which include an aromatic lipophilic end joined through an intermediate chain to a hydrophilic amine group. The molecules exist in an ionic or nonionic form, depending on the local pH. Local anesthetics penetrate into the cell membrane of the nerve, block the voltage-dependant sodium channels, and decrease the nervous conduction, avoiding the depolarization of the brain cell's membrane.

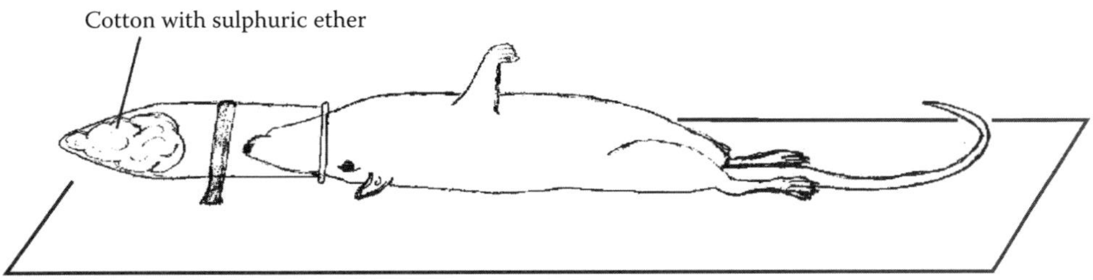

FIGURE 4.3 Maintenance of anesthesia with sulphuric ether.

There does not seem to be a great deal of specific information about the use of these agents in laboratory animals.

Local anesthetics should be used in surgeries where general anesthesia does not produce enough analgesia, for example, in thyroparathyroidectomy. For that, a volume of 50 μl of lidocaine chlorhydrate can be injected subcutaneously with a 15 × 0.5 injection needle in different parts of the surgery area when the rat is under the effect of a general anesthetic. Each injection should be given at 0.5 cm from the previous one.

MONITORING OF ANESTHESIA

Anesthesia is administered to block pain perception; therefore, the absence of response to the pain stimulus is essential. One of the criteria used to monitor the effect of anesthesia is the response of the animal to stimulus. The responses vary depending on the kind of anesthetic used, but, generally, what is evaluated is the presence or absence of different reflexes. The first reflex lost is rectitude, which consists of turning the animal back down and observing if it turns over chest down. The next reflex lost is swallowing or laryngeal. It is interesting to notice that, with ketamine, this reflex can be present, even with surgical levels of anesthesia. Another reflex is the palpebral; if it is present, the animal would blink. The most common reflex to determine if the animal is experiencing deep pain is the palmar reflex. The foot of the animal is extended and the skin between the toes is pricked. If it feels pain, it will take the foot away. Also when the ear is pricked, the rat will move the head or shake the ear if it feels pain. The same procedure can be made on the skin of the abdominal area.

PRECAUTIONS DURING ANESTHESIA

After the induction of the anesthesia, during the duration of surgery, it is necessary to keep the animal as near as possible to its normal physiological state. Three functions to be controlled include:

Room and body temperature: Although the temperature control is not one of the normal concerns of the investigator, especially when the rat is in a heated environment, it is a problem to consider during long surgeries. The mechanism to control body temperature is depressed during anesthesia. Animals lose a lot of heat through the body surface and the respiratory tract, causing a fast decrease in temperature that can reach 33°C in 20 minutes and less than 30°C in 1 hour. Hypothermia can generally cause the death of the animal. Therefore, the temperature must be monitored and maintained in the physiological range. The monitoring can be performed with an intrarectal thermometer. For that purpose, the maximum and minimum thermometers with internal and external sensors are very useful.

The maintenance of temperature must be performed with thermostatic boards (36 to 37°C). It can also be performed with incandescent heating lamps. An incandescent lamp of 100 watts at a 20-cm distance is enough if room temperature is around 20°C (see Chapter 44).

It is also important to consider that the presence of elements that raise the temperature locally will increase the loss of liquids through evaporation, especially when performing large incisions. An additional inconvenience is the dryness of mucosa and epithelium, due to temperature. This problem is solved with a saline solution that can be applied with a spray or externally with the aid of a piece of cotton.

Respiratory function: All anesthetics generally depress breathing. An easy way to monitor adequate respiratory function is by observing the color of the mucosa. The color of the plantar pads reveals if there is a severe hypoxia because it will turn cyanotic.

Cardiovascular function: It can be determined by pressing the mucosa. If, after pressing, it does not return to its normal color or if it is white, this means the animal is in shock.

Anesthesia Emergencies

Anesthesia emergencies are generally caused by human error. The causes may be the selection of the incorrect drug, the use of an inadequate dose, and/or the lack of experience to recognize and treat respiratory or circulatory failure before the animal reaches a state of shock.

To recognize heart failure, it is important to observe the color of the mucosas (white or cyanotic), the lack of beating in main arteries, lack of bleeding in the incisions, and undetectable heartbeats. In the case of a respiratory failure, it is also helpful to observe the color of the plantar pads.

Cardiopulmonary resuscitation can be done by pectoral massages, five compressions per artificial ventilation. The latter can be performed through the nostrils and can be done by insufflating through a pipette.

POSTSURGICAL CARE

The responsibility for the animal's well-being does not end once it leaves the surgery room. The postsurgery period consists of three stages:

1. Anesthesia recovery: It can be critical because most problems and physiological crises generally occur during this period. It requires frequent observation and care. An adequate recovery is one that allows, as soon as possible, the recovery of the physiological variables of the animal. The room where the animal recovers must be warm (27 to 30°C) and calm, to avoid stress that may produce an increase in the levels of corticoid and catecholamine, which can alter the results of the experiment.
2. Severe postsurgical care: The animal must be in the recovery area until it returns to normal condition.
3. Long-term postsurgical care: This happens when the animal returns to the physiological and behavior states that are as close to normalcy as possible. It is necessary to perform a series of routine monitoring procedures: remove sutures, return to normal motion functions, etc. An individual postsurgery register must be kept.

The key for good postsurgical care is a careful observation of the animal by trained staff. The frequency of the monitoring is determined by the nature of the surgical procedure and the recovery stage.

It is necessary to make sure that, during its recovery period, the animal will be protected from injuries caused by itself or other animals. It is important to observe if there are signs of infection or opening of the incision. A useful measure is the use of an Elizabethan collar to avoid the rat from removing the surgical stitches. It is important to make sure that, when using the collar, the animals will be able to access food and water.

REFERENCES

1. Flecknell, P. 1996. *Laboratory animal anaesthesia*, 2nd ed. London: Academic Press.
2. Van Dongen, J.J., R. Remic, J.V. Rensema, and G.H.J. Van Wunnik. 1990. *Manual of microsurgery on the laboratory rat*, Part I. Amsterdam: Elsevier Science Publishers B.V.
3. Kohn, D., S. Wikson, W. White, and G. Benson. 1997. *Anesthesia and analgesia in laboratory animals*. San Diego: Academic Press.
4. Mashour, G.A., S.A. Forman, and J.A. Campagna. 2005. Mechanisms of general anesthesia: From molecules to mind. *Best Pract. Res. Clin. Anaesthesiol.* 19(3): 349–64.
5. Xue, Q.S., and B.Q. Yu. 2003. The role of neuronal nicotinic acetylcholine receptors in the mechanisms of general anesthesia. *Sheng Li Ke Xue Jin Zhan*. 34(1): 37–41.
6. Hemmings, H.C. Jr, M.H. Akabas, P.A. Goldstein, J.R. Trudell, B.A. Orser, and N.L. Harrison. 2005. Emerging molecular mechanisms of general anesthetic action. *Trends Pharmacol. Sci.* 26(10): 503–510.

7. Urban, B.W. 2002. Current assessment of targets and theories of anaesthesia. *Br. J. Anaesth.* 89(1): 167–183.
8. Rudolph, U., and B. Antkowiak. 2004. Molecular and neuronal substrates for general anaesthetics. *Nat. Rev. Neurosci.* 5(9): 709–720.
9. MacDonald, E., and R. Virtanen. 1992. Review of the pharmacology of medetomidine and detomidine: Two chemically similar alfa 2-adrenoreceptor agonists used in veterinary sedatives. In *Animal Pain*, Eds. C.E. Short and A. Van Poznak. New York: Churchill Livingstone Publishers, pp. 181–191.
10. Suzdak, P.D., R.D. Schwartz, P. Skolnick, and S.M. Pauk. 1986. Ethanol stimulates gamma-aminobutyric acid receptor mediated chloride transport in rat brain synaptoneurosomes. *Proc. Nat. Acad. Sci. U.S.A.* 83: 4071–4075.
11. Maggi, C., and A. Meli. 1986. Suitability of urethane anesthesia for physiopharmacological investigations in various systems. Part 1: General considerations. *Experientia* 42: 109–114.
12. Krasowski, M.D., and N.L. Harrison. 1999. General anaesthetic actions on ligand-gated ion channels. *Cell Mol. Life Sci.* 55: 1278–1303.
13. Campagna, J.A., K.W. Miller, and S.A. Forman. 2003. Mechanisms of actions of inhaled anesthetics. *N. Eng. J. Med.* 348(21): 2110–2124.
14. Westphalen, R.I., and H.C. Hemmings Jr. 2006. Volatile aesthetic effects on glutamate versus GABA release from isolated rat cortical nerve terminals: Basal release. *J. Pharmacol. Exp. Ther.* 316(1): 208–215.
15. Villars, P.S., J.T. Kanusky, and T.B. Dougherty. 2004. Stunning the neural nexus: Mechanisms of general anesthesia. *AANA J.* 72(3): 197–205.

5 Euthanasia

Alfredo Rigalli

INTRODUCTION

Euthanasia is the sacrifice of an animal in keeping with the ethical principles. As a consequence, it can be considered an important part of the experiments. An extensive revision of the methods, as well as the ethical aspects, are available in numerous publications .[1-4]

CAUSES FOR EUTHANASIA

1. To avoid pain and suffering during the experiment.
2. Because it is not possible to use the animal in another experiment.
3. To obtain biological samples in an invasive way, e.g., the liver or femur.

Unnecessary animal sacrifice reveals an irresponsible attitude of the investigator toward the funds destined for investigation as well as a lack of respect for the animals.

Requirements to Perform Euthanasia

1. The process must be fast.
2. The process must avoid fear, pain, and suffering.
3. The process must be performed in an unconscious animal.
4. Euthanasia must be performed when the animal is under lack of sensitivity.
5. It must cause heart and breathing failure.
6. The surgeon or the member of the staff must euthanize the rat working professionally and respectfully. To do that, they must know:
   a. How to recognize the signs of fear, anxiety, pain, and suffering in the animal.
   b. The correct use of the equipment for the surgery and euthanasia.
   c. How to produce unconsciousness, analgesia, and anesthesia.
   d. How to recognize the death of the rat.
   e. The steps to euthanize a rat.
   f. The maintenance of the equipment.
   g. How to do everything in order to teach others about good practice in euthanasia.

Acceptable Techniques

1. Inhalation of volatile anesthetics: Volatile anesthetics, such as halothane, isoflurane, or sulphuric ether can be used. Halothane is advisable because it is fast and a nonirritant to the animal. It has the disadvantage of being accidentally inhaled by humans. As a consequence, a mechanism for gas excess elimination is needed.
2. Injectable anesthetics: An overdose of any injectable anesthetic is acceptable. Ketamine or pentobarbital can be used as injectable anesthetics. They produce fast unconsciousness induction and they are cheap. The disadvantage is that the drug remains in the animal once it is dead.

3. Intravenous injection of KCl (potassium chloride) or $MgSO_4$ (magnesium sulfate): It is only acceptable after inducing anesthesia with other anesthetics. As advantages, they are cheap and reduce the anesthetics expenses. KCl or $MgSO_4$ can be administered by catheterization or intracardiac injection.
4. Cervical fracture: It is only acceptable when the animal is under profound effects of anesthetics. It is fast, without costs, it does not require specific equipment and avoids contamination by drugs. As disadvantages, it is aesthetically unpleasant and needs special training. It is not a recommended way to euthanize a rat.
5. Decapitation: It is only acceptable when the animal is under anesthetic effects. As advantages, it is fast, without costs, and avoids contamination by drugs. The disadvantages are: special training is needed, it is aesthetically unpleasant, and it is only applicable on small animals. As an additional disadvantage, it requires a guillotine.
6. Blowing the head: It is only acceptable under profound anesthetic effects. It is fast, without cost, it does not require specific equipment, and it avoids contamination by drugs. However, it needs special training, it is aesthetically unpleasant, and it is only applicable on small animals.

Process Recommended on Rats under Anesthesia

The intracardiac injection of 0.5 ml saturated KCl/100 g of body weight produces sudden death. KCl induces cardiorespiratory failure in 1 to 2 seconds. It can produce slight movements; however, convulsions or contractions are not usually produced; it causes the immediate relaxation of the animal. A disadvantage is that it requires practice to perform the heart puncture. If the rat is catheterized, the injection of the solution can be performed through the catheter in a vein or artery. However, vein catherization is preferable.

REFERENCES

1. Canadian Council on Animal Care Guidelines. 1993. Euthanasia. In *Guide to the care and use of experimental animals*. Vol 1, Chap. 12, 2nd ed.: www.ccac.ca
2. European Commission. 1996. Recommendation for euthanasia of experimental animals. Part 1. *Laboratory Animals* 30: 293–316: www.lal.org.uk/pdffiles/la1.pdf
3. American Veterinary Medical Association (AVMA). 2007. AVMA guidelines on euthanasia. www.avma.org/resources/euthanasia.pdf
4. European Commission. 1997. Recommendation for euthanasia of experimental animals. Part 2. *Laboratory Animals* 31: 1–32: www.lal.org.uk/pdffiles/la2.pdf

6 Antibiotics

Verónica E. Di Loreto and Alfredo Rigalli

INTRODUCTION

The resistance of the rat in contracting infections after a surgery is well known, except in those cases where intestinal content contamination occurs. In this case, death caused by peritonitis occurs in 72 hours in 100% of the cases.[1]

Information about the use of antibiotics is somewhat lacking and, in many cases, the extrapolation of the doses and antibiotics used on other species is necessary in helping us correctly perform our surgeries. In our case, the most common experience is the use of antibiotics during abdominal surgeries, especially during those surgeries where there is contact with the intestinal content.

Ceftriaxone, a third-generation cephalosporin, has been used with results similar to those reported by other investigations.[2,3] Parenteral administration together with peritoneal irrigation gives better results than the injectable administration used as the only route of administration.

During abdominal surgery, it is intramuscularly injected after anesthesia is induced: 3 mg ceftriaxone/100 g of body weight (0.06 ml of 50 mg/ml ceftriaxone per 100 g of body weight). Before suturing the abdominal incision, irrigation with 10 ml of 10 mg/l ceftriaxone solution must be performed, removing the excess with a vacuum pump. The intramuscular dose is then repeated every 24 hours for 3 days.

In surgeries not involving the abdominal cavity, infection is not common, but penicillin treatment is recommended as a precaution.[4]

REFERENCES

1. Perdue, P.W., K.K. Kazarian, J. Nevola, W.R. Law, and T. Williams. 1994. The use of local and systemic antibiotics in rat fecal peritonitis. *J. Surg. Res.* 57(3): 360–365.
2. Stewart, D.J., and N.A. Matheson. 1978. Peritoneal lavage in faecal peritonitis in the rat. *Br. J. Surg.* 65(1): 57–59.
3. Eyal, A., F. Greif, I. Shalit, and S. Lelcuk. 1996. Acute bacterial peritonitis: Permeability of cephalosporins in the peritoneal cavity. *Isr. J. Med. Sci.* 32(12): 1317–1319.
4. Patzakis, M.J., L.D. Dorr, W. Hammond, and D. Ivler. 1978. The effect of antibiotics, primary and secondary closure on clostridial contaminated open fracture wounds in rats. *J. Trauma.* 18(1): 34–37.

Section II

General Procedures

7 Operating Theatre

Verónica E. Di Loreto and Alfredo Rigalli

INTRODUCTION

To work with rats, the operating theatre must be, first of all, comfortable for the surgeon. It must have an operating table, ample light, and adequate optical magnification elements. The basic materials for surgery on rats include:

1. Thermostatized rat board (surgical table, stretcher): It can be a 20 × 30 cm plastic board to which the animal can be held with sticking plaster.
2. Light: Portable if possible or with a mobile base. It can be incandescent (100 W) and can be useful as a heat source as well.
3. Magnifying glass with light: Preferably a cold light (round fluorescent tube).
4. Protection glasses: Made of polycarbonate. Can be bought in surgical or industrial safety shops.
5. Disposable latex gloves: It is essential that they fit the hand perfectly to make the surgical practices easier. Available in large, medium, small, and extra small sizes.
6. Surgical mask.
7. Surgical gown.
8. Cotton balls of 1 cm in diameter.
9. Heparin.
10. Saline solution (NaCl 9 g/l) disposable bag.
11. General anesthetic: Appropriate to each technique.
12. Local anesthetic.
13. Syringes: 1, 3, 5, and 10 ml (Figure 7.1).
14. Injection needles: The length and the diameter are expressed in millimeters, e.g., 25 × 0.8, which means 25 mm in length and 0.8 mm of outer diameter (Figure 7.1). The number can also be written as 25 × 8, where the first number states the length in millimeters and the second indicates the outer diameter in tenths of millimeters. The size of a needle can also be written in a nonmetric scale, e.g., 21G × 1". 21 G is the outer diameter (the greater the first number, the lower the diameter). The second number expresses the length in inches. The most common injection needles for surgery in the rat are:
    - 14.1 13 × 0.4 (13 × 4 or 27G × 1/2"): For local anesthesia, intramuscular injection (i.m.), subcutaneous injection (s.c.), and intraperitoneal injection (i.p.) in rats weighing less than 50 g; medical use: subcutaneous injection.
    - 14.2 25 × 0.8 (25 × 8 or 21G × 1"): For heart puncture and intraperitoneal injection; medical use: vein blood extraction.
    - 14.3 13 × 0.3 (13 × 3 or 30G × 1/2"): For intravenous injection through the vein of the tail; medical use: intradermic injection.
15. Container for discarding injection needles. It must be made of plastic with a hole to place the needle after use.
16. Catheters (Table 7.1).
17. Surgery materials.
    In each chapter, synonymous or equivalent materials and their amounts are shown in brackets.

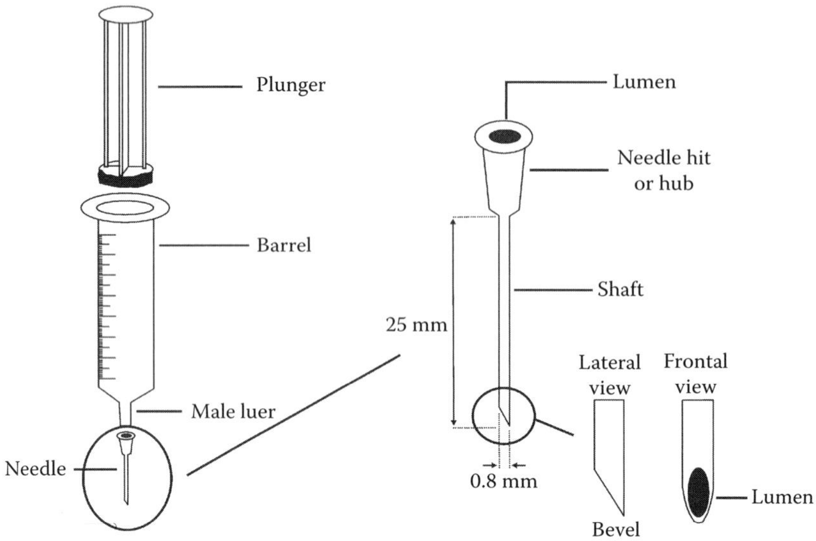

FIGURE 7.1 Parts of a conventional syringe and needle.

17.1 Ring-handled preparation scissors, curved or straight, blunt/blunt (B/B), or blunt/sharp (B/S) (Figure 7.2).
17.2 Ring-handled iris scissors, straight or curved, sharp/sharp (S/S) (Figure 7.3).
17.3. Adson forceps (large, straight surgical forceps, straight dissection forceps).
17.4 Rat-tooth forceps (large, straight surgical forceps).
17.5 Microforceps (jewelry forceps) (Figure 7.4).
17.6 Cotton pliers (curved forceps) (Figure 7.5).
17.7 Straight suture needle.
17.8 Curved suture needles described in the book by their curvature and diameter, e.g., suture needle curved ½ – 5 – 10 mm means one half of circumference and 5 to 10 mm of diameter.
17.9 Thread.
17.10 Weitlaner–Locktite retractor.
17.11 Kocher forceps.
17.12 Artery forceps (baby mosquito, Halsted forceps) (Figure 7.6).
17.13 Spatula: The odontological one, with a plain rectangular end and a plain triangular end, is the most useful (Figure 7.7).
17.14 Iridectomy scissors 45° angled: Only used in surgeries with a very small surgical area.
17.15 Needle holder (needle driver, Mayo needle holder, Heager needle holder) (Figure 7.8).

TABLE 7.1
Description of Catheters

Catheter	Medical Use	Use for Surgery on Rats	Outer Diameter/ Inner Diameter (mm)
PC-10		• Ureter catheterization (PC-10 or PC-40)	0.4/0.3
PC-40 PC-50/B-TC-50 or similar	• Polyethylene tubes for vein cannulation and epidural	• Artery catheterization: celiac, mesenteric and femoral artery • Vein catheterization: femoral, portal vein	0.5/0.4
PC-75/B TC75 PC-100 or similar	• Polyethylene tubes for vein cannulation and epidural	• Aorta and vein cava catheterization	1.3/1.1
BSN35 K-35 or similar	• Catheter for intragastric feeding of premature newborn baby	• Tracheostomy • Orogastric catheterization, especially for rats with a body weight lower than 100 g • Bladder catheterization • Aorta artery and vein cava catheterization	1.4/1.1
K-31 or similar	• Catheter for intragastric feeding of children	• Orogastric catheterization, especially for rats of body weight higher than 100 g • Stomach isolation *in situ* • Intestine isolation *in situ*	2/1.4

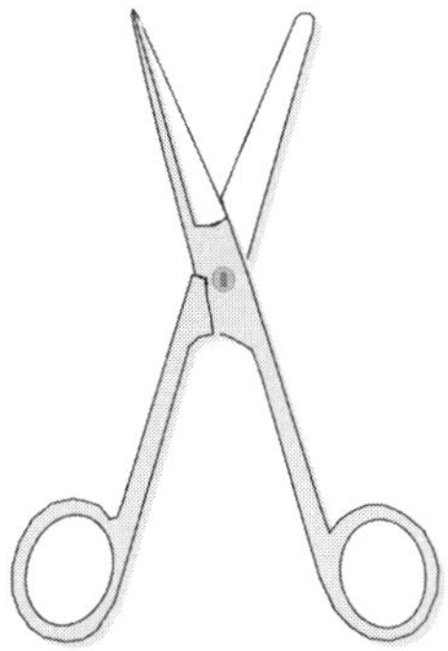

FIGURE 7.2 Ring-handled preparation scissors, straight, blunt/sharp.

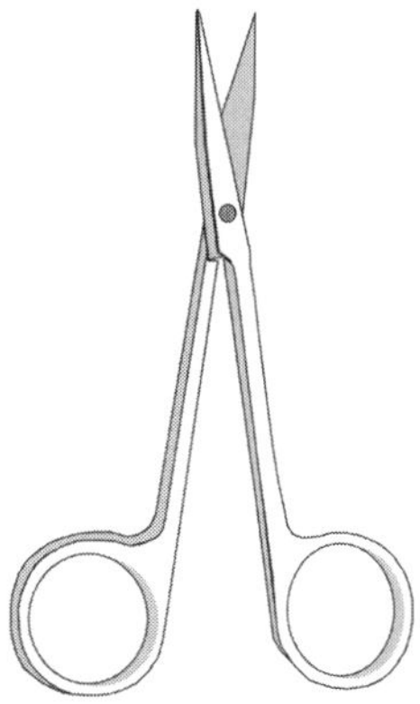

FIGURE 7.3 Ring-handled iris scissors, straight, sharp/sharp.

FIGURE 7.4 Microforceps.

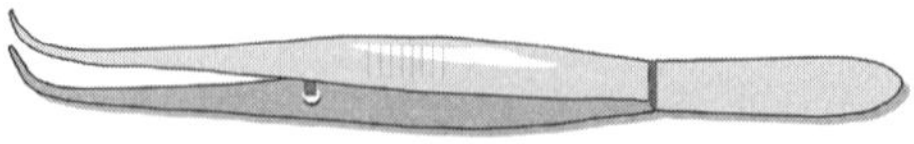

FIGURE 7.5 Cotton pliers.

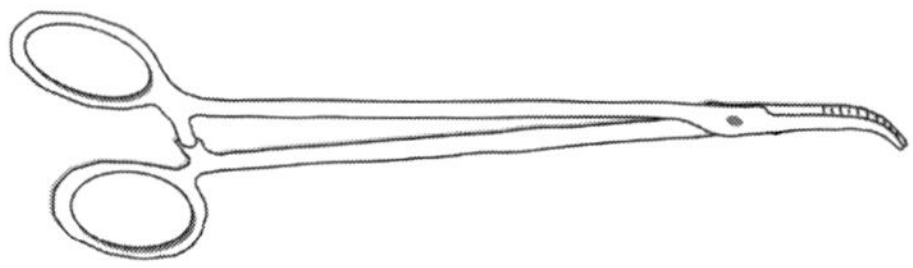

FIGURE 7.6 Artery forceps.

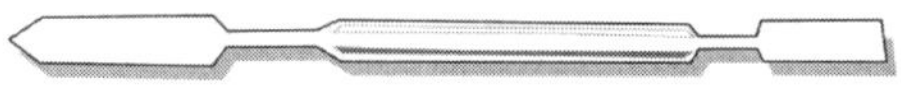

FIGURE 7.7 Odontological spatula.

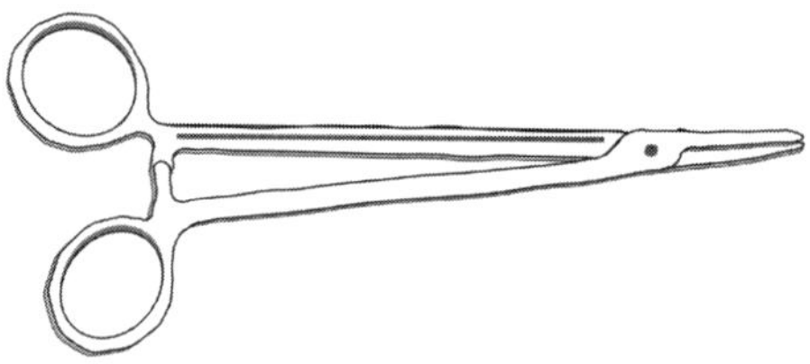

FIGURE 7.8 Needle holder.

8 Rat Identification

Verónica E. Di Loreto and Alfredo Rigalli

INTRODUCTION

Generally, while experimenting with rats, it is necessary to identify each animal. With this aim in mind, rats are marked with nicks on the ears when weaning.

A diagram of these nicks and the corresponding numbers of each nick are shown in Figure 8.1, which shows the animal from the rear view. Figure 8.2 shows a full-face image of the animal. It depends on each person's preference which position he or she chooses at the moment of reading the number.

Nick numbers that correspond to 30, 10, 3, and 1 can be found three times more often on animals. The 800, 400, 200, and 100 nick numbers can be found only once. To know the number of the rat, the different nick numbers on the ears must be added. Another important detail in identification is the fact that female rats have even numbers, and male rats have odd numbers.

Figure 8.3 and Figure 8.4 are examples of the procedure. In the first case, the number of the animal is 596; therefore, it is a female rat. In the second case, the number of the rat is 975, thus it is a male rat.

If the nicks on the ears cause confusion identifying two or more animals, painting the back of one of them with a solution of methylene blue 5% w/v or with silver nitrate ($AgNO_3$) 1% w/v will make recognition easier. The dye lasts 1 to 2 weeks and it does not interfere with the experiment.

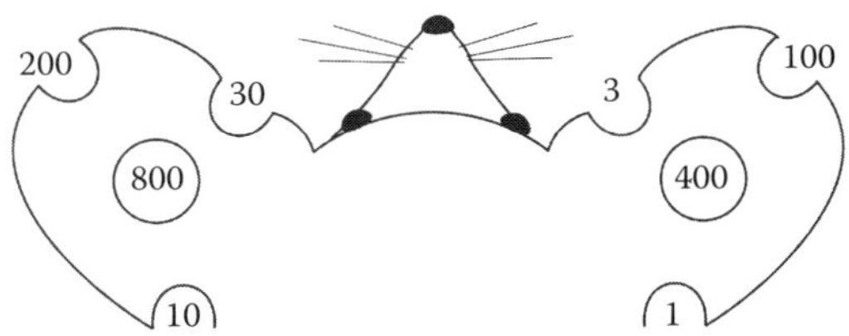

FIGURE 8.1 Nicks on the rat ears seen from behind.

FIGURE 8.2 Nicks on the rat ears seen from the frontal full face.

FIGURE 8.3 The nicks represent number 596.

FIGURE 8.4 The nicks represent number 975.

9 Antisepsis, Sterilization, and Asepsis

Diego Holotte, Jésica Nipoti, Maria J. Pretini, and Verónica E. Di Loreto

INTRODUCTION

Infections are rare in the rat, even after a surgery. However, an infection can cause unnecessary suffering and death of the animal and modifications in the results of the experiment. A brief description of the common tools for a safe surgical area is given in this chapter.

DISINFECTION AND ANTISEPSIS

Disinfection is the destruction of pathological organisms, substances, and inorganic materials of the environment. Detergents, ethanol 70% w/v, and sodium hypochlorite 1% w/v are the most common and effective; they are also low-cost products. However, they cannot be used for antisepsis because they are skin irritants.

Antisepsis is the elimination or decrease in the number of microorganisms in organic surfaces. Iodopovidone, chlorhexidine, and hydrogen peroxide are effective and have a less harmful effect on the skin. Because of the higher cost, they are not used as disinfectants.

STERILIZATION

Sterilization is a method to eliminate all the microorganisms, including spores, from surfaces and objects. The method used can be chemical or physical.

Chemical Sterilization:

1. Gas sterilization: ethylene oxide, formaldehyde, and betapropiolactone are used.
2. Organic chemical agents: glutaraldehyde is a powerful bactericidal and sporicide.

Physical Sterilization:

1. Wet heat: vapor under pressure. For this, an autoclave or a pressure cooker is necessary.
2. Dry heat: an electrical heat cabinet or an infrared radiation oven is necessary.
3. Ultraviolet radiation: mobile or fixed lamp.

ASEPSIS

Asepsis is the methods or procedures used to preserve sterility.

GENERAL CONSIDERATION FOR THE OPERATING THEATER

It is advisable for the operating theater to be a closed place, with the least amount of furniture possible. The floor, walls, and ceiling are constructed of material that is easy to wash.

Any person entering the theater who is not part of the surgical procedure is a threat to asepsis and this must be avoided.

Presurgery Procedures

Before performing the procedures described below, the material for the scheduled surgical process must be selected.

The materials to be sterilized through wet heat include: cotton, thread, sanitary pads, gloves, the surgical field, white coats, caps, surgical masks, and gauzes. These items must be perfectly clean as bacteria in dried blood and dirt inhibits the sterilization action of wet heat. Surgical materials should be packed in two manila paper envelopes (that we will call external and internal) and placed inside the autoclave. The exposure period is approximately 20 minutes depending on how the temperature and pressure are adjusted. In case this equipment is not available, a pressure cooker can be used. The material is placed in the cooker inside a container with holes, which allows the entrance of vapor, but avoids direct contact with water. In the cooker, sterilization is achieved 30 minutes after the water reaches the boiling point. Sterility will last for two weeks. After this time, materials must begin a new sterilization cycle.

The materials designated to undergo dry heat sterilization include: metallic surgical equipment (scissors, suture needles, forceps) and the thermo-resistant glass elements. They must be washed with water and detergent, meticulously brushed and dried, then placed in the heat cabinet for 1 hour at 160 to 180°C. A tape that changes color when the appropriate temperature and pressure are reached is not an adequate guide for confirming the sterilization of the material.

The elements that do not withstand high temperatures, such as syringes, catheters, etc., must be sterilized through chemical means. The substances used for this procedure are expensive, highly toxic, and flammable. For the cost conscious, this material can be found in the market already sterile and ready to use.

The operating theater and its devices should be disinfected with sodium hypochlorite if possible, the day before the surgery. Ultraviolet light sterilizes air and surfaces. Do not use this when the staff is present because excessive exposure damages the skin.

As a conclusion, the preparation of the material will be accomplished if the following steps are performed: decontamination, cleaning, rinsing and drying, classification, assembly, packaging, sterilization, control, storage, transportation, and duration of sterilization.

Intraoperating Procedures

It is recommended that at least two persons work on intraoperating procedures, an assistant and a surgeon. The former will organize the operating theater, dress the surgeon, and prepare the animal. In this way, the surgeon will be able to focus his attention fully on the surgery and the antiseptic measures.

To begin with, the assistant must wash his hands meticulously with a brush and antiseptic soap and put on sterile clothing. Then, he must cover the operating board with a sterile surgical field. The surgical material must be placed within reach of the surgeon. The assistant removes the external envelope at the beginning of surgery. The sterile internal envelope can be manipulated directly by the surgeon.

The surgeon must wash his hands in the same manner as the assistant. With the aid of the assistant, he must put on sterile clothing, such as white coat, cap, surgery mask, boots, and gloves. The surgeon must remain inside the operating theater without making contact with any surface up to the moment of the start of the surgery.

The preparation of the animal begins with disinfecting the surgery board with alcohol. The incision area of the animal is shaved and painted with iodopovidone with a spiral movement toward the periphery. At this point, the assistant must change his gloves because they have become contaminated by the manipulation of the animal. Finally, he must cover the working area with a sterile fenestrated drape to avoid the contact of the surgeon and the instruments with the animal and the surgery board.

The pathological waste must be eliminated in red bags and the sharp objects placed in the discard container.

Although it is not possible at times to accomplish all the proposed measures, it is important not to become frustrated, as it is always better than doing nothing.

10 Suture

Aneley Traverso and Alfredo Rigalli

INTRODUCTION

The suture of tissues, a fundamental moment in the surgery, must be handled skillfully by the surgeon. It is extremely useful to anastomose blood vessels and to keep the tissues in apposition until cicatrisation (formation of scar tissue).

Nowadays, there are several materials and techniques for suturing that change and improve constantly. However, considering its particular application in experimental animals, only the most common technique will be described.

MATERIALS

Needles

Made of tempered stainless steel, the needle must pass through the tissues causing the least amount of damage possible.

The needle consists of three elements: eye, body, and end. According to the body, the needles can be classified longitudinally as straight or curved, and by the transverse section of its body as circular, rectangular, or triangular. The end of the needle is either tapered (cylindrical) or cutting (two opposed edges). The cutting end is used in tissues that are thicker, harder, or more fibrous (Figure 10.1).

Curved needles can by classified by their curvature as 1/4, 3/8, 1/2, 5/8 of a circle (Figure 10.2), and by their diameter. The most common in general surgery is the ½ of curvature and 10 to 20 mm of diameter needle (see Figure 10.1).

Needle Holder

They are necessary for the handling of curved needles (see Chapter 7).

Forceps

They are necessary to handle the tissue. Adson forceps and the same forceps with rat tooth are the most commonly used.

Suture Thread

Classification:

1. *According to the number of filaments*:
   Monofilament thread: They are better tolerated, but are more expensive and more difficult to handle; examples: nylon, wire.
   Multifilament thread: They are cheaper and easier to handle; example: silk, cotton, polyester, linen.
2. *According to absorbability*:
   Nonabsorbable: They must be removed from the tissue after cicatrisation; examples: silk, cotton, nylon, linen.

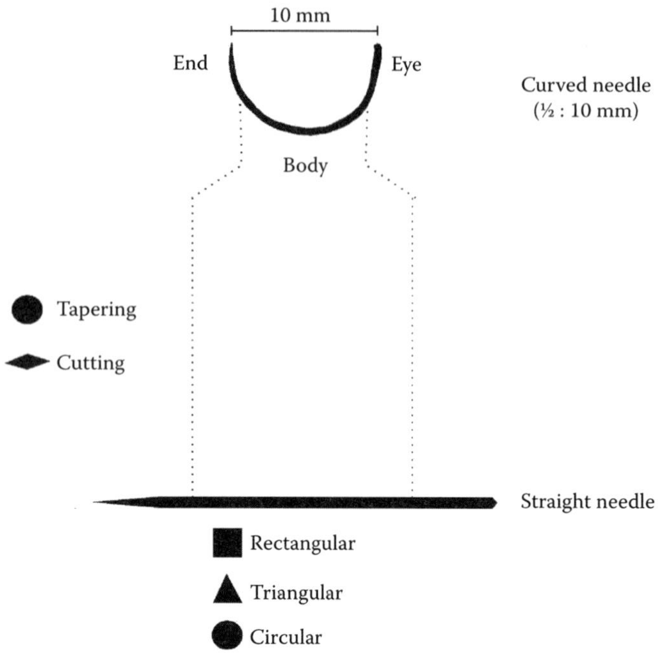

FIGURE 10.1 Elements and classification of needles.

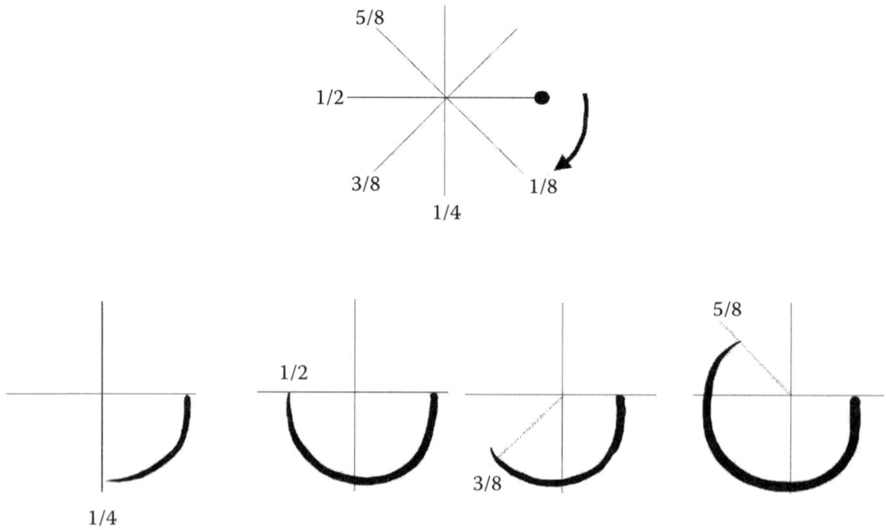

FIGURE 10.2 Curvature of needles.

Absorbable: Most of them lose their tension in 60 days, although the degradation is not complete; examples: simple catgut, chromic catgut, polyglycolic acid.

3. *According to its origins*:
   Natural: simple catgut, chromic catgut, cotton, silk, linen
   Synthetic: polyglycolic acid, nylon, polyester.
4. *According to its size*:
   The size of the thread is indicated with a number. It can be a natural number (1, 2, ...9) or a number followed by zero (1/0 or 10, 2/0 or 20 ..... 10/0 or 100). As regards natural numbers, the bigger the number, the greater the thickness; example: number 3 has a higher size than number 2. In a number followed by zero, the greater the number of zeros, the smaller the thickness of the thread. For example: 2/0 or 20 thread has a greater size than 3/0 or 30.

Generally, the use of many of the previously mentioned threads is determined by the cost/benefit ratio and the surgery. Therefore, when choosing the materials, it is important to give priority to the following characteristics: inexpensive, sterile, multipurpose, and keeps its structure after being used. Considering all of these premises, linen and nylon are the most used threads.

Surgical Sutures

Initial Considerations

Needle holders are necessary when using curved needles because they allow prone–supination movements, introducing the needle in a right angle according to the tissue. An adequate pressure should be applied, avoiding ischemia of the suture area. The least amount of stitches possible must be done, which varies according to the resistance of the tissue.

Simple Stitch and Knot Technique

A needle holder and an Adson forceps are needed to perform the stitch. The following description of the technique is for a right-handed surgeon, but it can be adapted for a left-hander.

The curved needle is held with the needle holder in the right hand. The needle is introduced in a perpendicular direction to the tissue (Figure 10.3a). The distance between the point where the needle is introduced and the incision must be half of the distance between two consecutive stitches.

With a supination movement, the end of the needle must pass both edges of the incision, perpendicular to the incision (Figure 10.3b). Then, the end of the needle is held with the needle holder (Figure 10.3c) and, with a supination movement, it is extracted from the tissue (Figure 10.3d). The thread is passed through the incision until a short part of it (2 cm) remains in the opposite side of the incision (Figure 10.3e). One loop is made over the Adson forceps in a clockwise movement (Figure 10.3f)

Then, the right end of the thread is held with the Adson forceps (Figure 10.4a) and movements to the right with the needle holder and to the left with the Adson forceps are performed (Figure 10.4b/c).

The end of the thread with the needle is cut leaving a 5 cm end. The left end of the thread is held again with the needle holder. One loop is made over the Adson forceps in two opposite directions (Figure 10.5).

If loops are made in a counterclockwise direction, a parallel square knot is obtained (Figure 10.6a). If the loops are made in the clockwise direction, a crossed loose knot is obtained (Figure 10.6b). The former is preferred for suture in the rat.

In Figure 10.6, both knots are represented. In the left, the final position of threads is shown for the parallel square knot and, in the right, for the crossed loose knot. A schematic representation of a stitch is also shown (Figure 10.6c). The vertical continuous line represents the line of

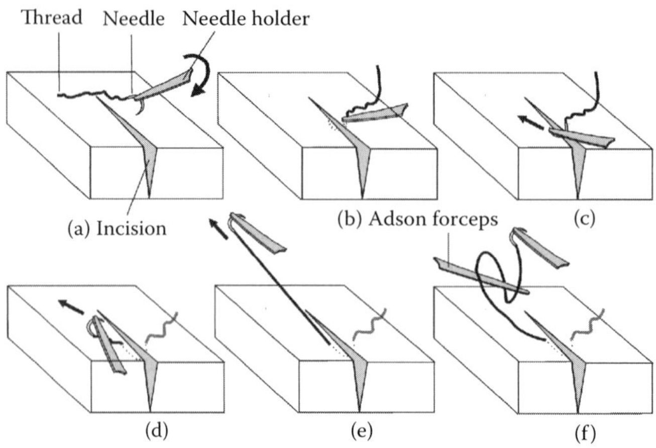

FIGURE 10.3 Demonstration of a knot technique.

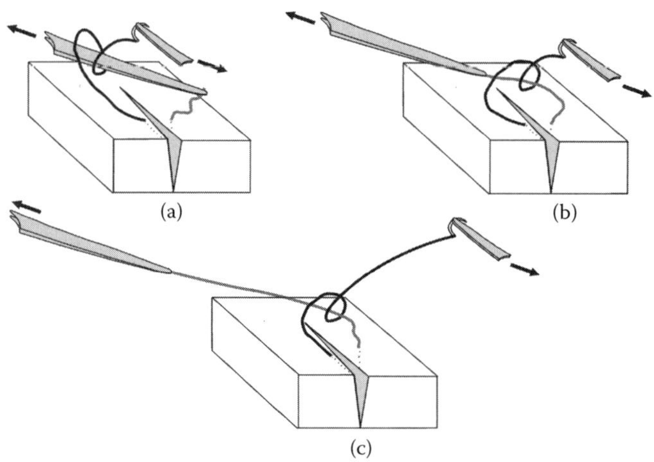

FIGURE 10.4 Demonstration of a knot technique.

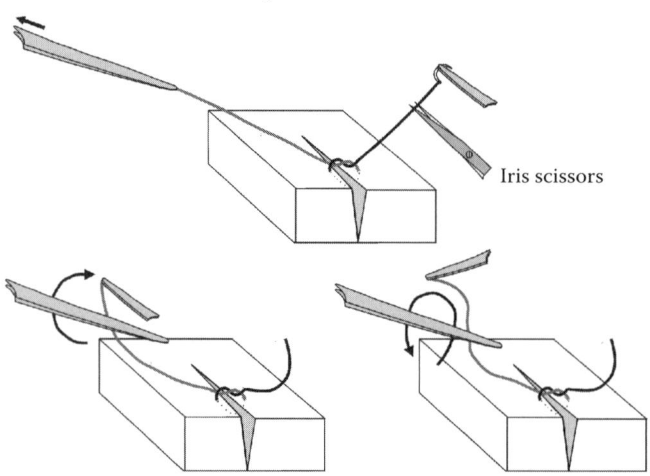

FIGURE 10.5 Demonstration of a knot technique.

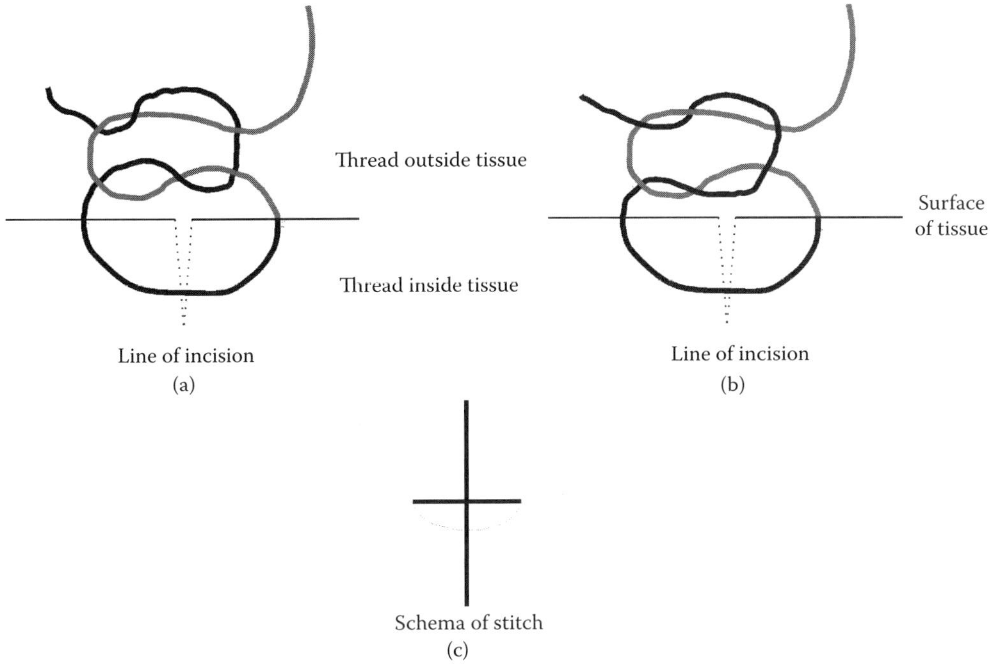

FIGURE 10.6 Position of threads in a parallel square knot (top left) and in a crossed loose knot (top right). Black and gray are used to distinguish both ends of the thread. Schematic representation of the stitch (bottom): the vertical continuous line, the horizontal continuous line, and the dashed line represent the incision, the thread outside the tissue, and thread inside the tissue, respectively.

incision; the straight horizontal line, the thread on the incision; and the dashed line, the thread inside the tissue.

Once the stitch is finished, the knot must be moved to one side of the incision.

Types of Sutures

Interrupted

Used when most tensile strength is needed. Simple stitches are done uniformly separated through the edges of the incision as described previously. They are the most used and the easiest to remove. The most common interrupted sutures are simple stitch, vertical mattress suture, and horizontal mattress suture.

Simple Stitch

It is the most used because it is faster and more practical (Figure 10.7).

Vertical Mattress Suture

It allows one to suture in different levels and to obtain a good apposition of the incision edges (Figure 10.8).

Horizontal Mattress Suture

It is used in incisions where there is excessive tension. It prevents dehiscence and has good apposition of the incision edges (Figure 10.9).

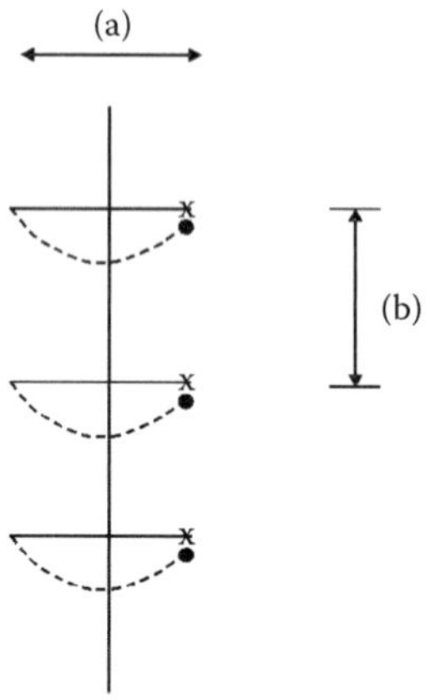

FIGURE 10.7 Simple stitch. The vertical line represents the incision, the horizontal lines, the thread outside the tissue, and the curved dashed lines, the thread inside the tissue. The points indicate the places where the needle is introduced and the crosses indicate the places where the knots are made. The distance (a) should be equal to (b).

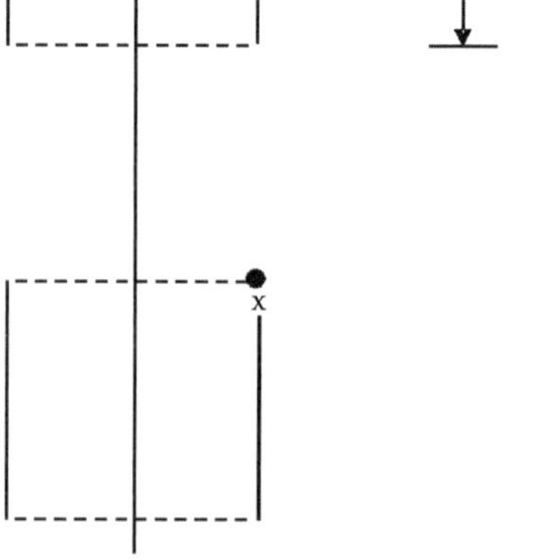

FIGURE 10.8 Vertical mattress suture. The vertical line represents the incision, the horizontal dashed lines, the thread inside the tissue, and the continuous vertical segments, the thread outside the tissue. The points indicate the places where the needle is introduced and the crosses indicate the places where the knots are made. The distance (a) should be equal to (b).

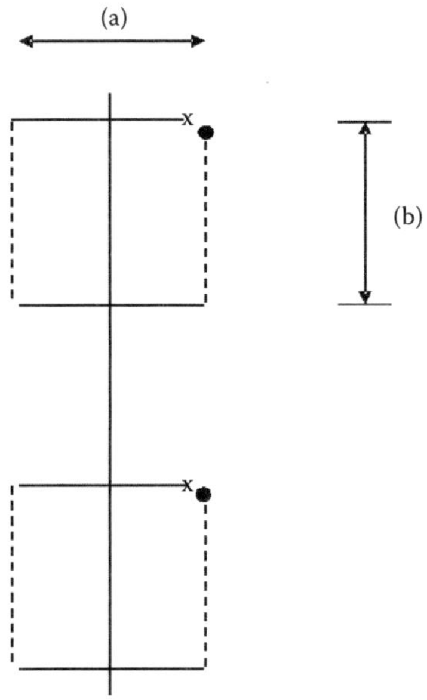

FIGURE 10.9 Horizontal mattress suture. The vertical line represents the incision, the horizontal lines, the thread outside the tissue, and the vertical dashed segments, the thread inside the tissue. The points indicate the places where the needle is introduced and the crosses indicate the places where the knots are made. The distance (a) should be equal to (b).

Continual

In a continual suture, it is more difficult to remove the stitches and, as a consequence, it is recommended that one use absorbable thread. The continual sutures are faster and sometimes used in muscles (Figure 10.10). If there is any suspicion of infection, they are not used.

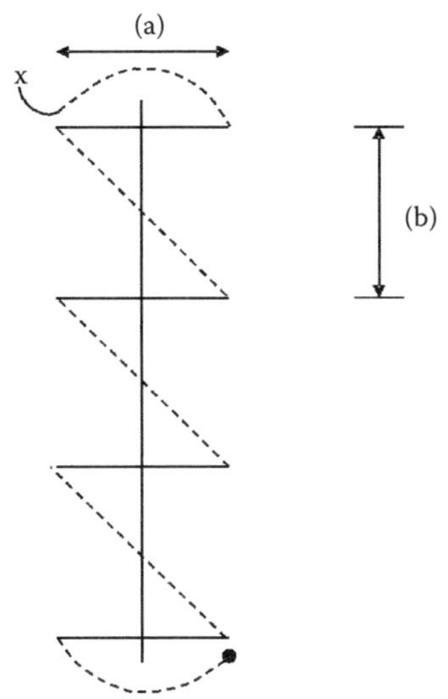

FIGURE 10.10 Continual suture. The vertical line represents the incision, the horizontal lines, the thread outside the tissue, and the dashed segments, the thread inside the tissue. The point indicates the place where the needle is introduced and the cross indicates the place where the knot is made. The distance (a) should be equal to (b).

11 Substances Administration

Maela Lupo, Verónica E. Di Loreto, and Alfredo Rigalli

INTRODUCTION

The routes for the administration of substances include intravenous (i.v.), intraperitoneal (i.p.), intramuscular (i.m.), subcutaneous (s.c.), intragastric (i.g.), inhalatory, and through drinking of water. These routes allow the administration of exact amounts of substances in solution or suspension. The inhalatory route is the exception, and it is useful in the case of gaseous and volatile substances (anesthetics, hydrogen, oxygen, etc.).

In all cases it is necessary to take into account certain practical considerations:

1. Water-soluble drugs and compounds must be dissolved in distilled water. If necessary, sterile saline solution can be used.
2. Insoluble compounds can be administered in suspension as long as they go through a subsequent dissolving process. For example, calcium carbonate is insoluble, but when it is orally administered, it is dissolved in the acid medium of the gastric lumen.
3. The volume is determined by the route of administration. For a 200 g average weight rat, the approximate maximum volumes would be: i.v.: 1 ml, i.p.: 5 ml, i.m.: 0.25 ml, s.c.: 0.5 ml, and i.g.: 10 ml (depending on the gvastric capacity of the rat).
4. pH of the solution: The control of this variable will depend on the route of administration. When using intravenous, intramuscular, subcutaneous, or intraperitoneal routes, it is preferable to use solutions with a pH based on the physiological range even if the rats can tolerate a wide range of pH (5.4 to 8). In case this does not happen, the solution must be injected slowly to allow the blood to buffer the solution. In the intragastric administration, the variation margins may be larger. It is important to understand that the gastric content in fasted rats is neutral; however, it can tolerate extremely acid solutions.
5. Rate of injection: Generally, all the injections must be administered slowly. Intraperitoneal, subcutaneous, and intragastric routes can be performed more quickly. The factors that can influence the flow are the physiological effects of the compound, its concentration, the half-life (e.g., alloxan, a drug used for experimental diabetes induction, must be injected quickly because it is very unstable and easily inactivated), solution pH, and maximum blood concentration.
6. Temperature of the solution: If possible, the fluids must be injected at body temperature.

ADMINISTRATION IN DRINKING WATER

The administration of substances in drinking water is not very useful in cases where the dose must be strictly controlled. However, this way is useful in certain experiments and the amount of substance administered subsequently can be calculated through water intake. It must be taken into consideration that water intake increases, but water intake relative to body weight decreases as the rat grows (Figure 11.1). As a consequence, if a constant dose must be administered, the concentration of the drug in the solution should be reduced throughout the experiment. In contrast, if a definite amount of substance per gram of body weight is needed, the concentration of the drug should be increased. Despite the previous corrections in the concentration as the rat grows, the dose will not be exact because of the variance in the volume of water the rat drinks (Figure 11.1).

INJECTABLE ROUTES

The injectable route must be chosen in accordance with the volume to be injected and the biological effect of the substance. The greater the rate of absorption and the rate at which it reaches the blood stream, the lower the dose to be injected. When more than one route can be selected, the least stressful for the animal must be used. Independent of the route of injection, the area must be thoroughly disinfected with ethylic alcohol.

As a reference, the rate to reach the blood stream is listed in a decreasing order: intravenous, intraperitoneal, intramuscular, subcutaneous, and, finally, intradermic. Apart from the intravenous injection where the substance is injected into the blood stream, in the other routes, the substance must be transported through one or more tissular structures. As a consequence, the rate of absorption

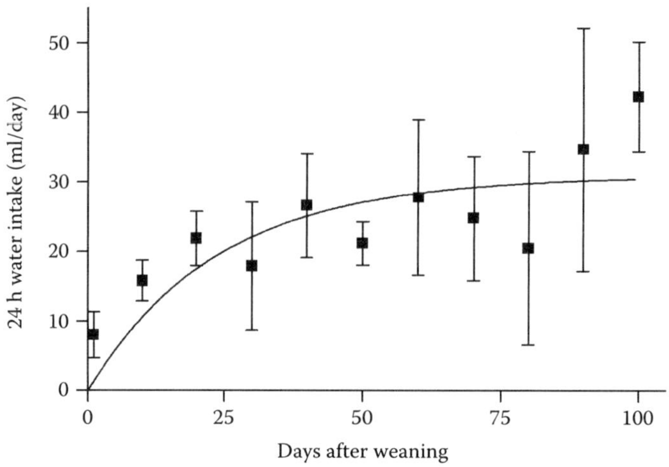

24 h water intake = 30.86 * (1−exp(−0.042* days after weaning)

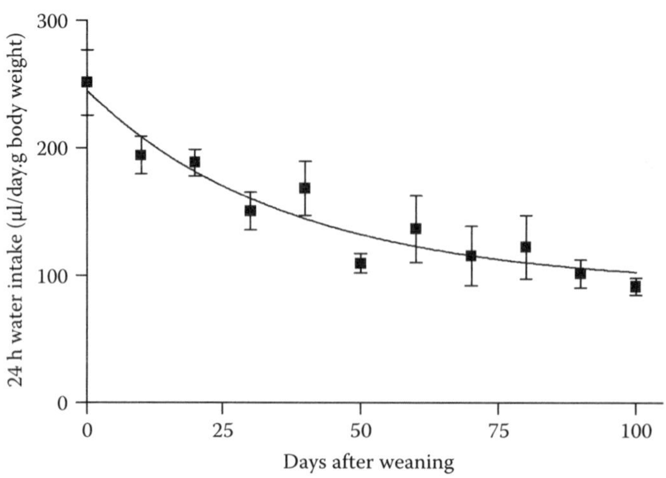

24 h water intake = 152.3 * exp(−0.027* days after weaning) +92.2

FIGURE 11.1 Shown is 24-hour water intake (top) and 24-hour water intake/body weight (bottom) for 100 days after weaning. Equations that fit the values are shown at the bottom of each figure.

is also influenced by the chemical properties and the molecular weight of the substance. As a general rule, the rate of absorption has an indirect relationship with the hydrophilic property and the molecular weight of the substance.

Subcutaneous Injection

This method of injection is needed when small volumes of solution or very active substances must be administered. The rate of absorption is lower than in the other routes of injection and the effect of the substance is more sustained than in the intravenous or intraperitoneal injection. In contrast, the dose of the substance must be higher than for the other mentioned routes. It is useful for the administration of analgesic and some active substances, such us adrenalin. The maximum volume for a 200-g rat is 0.5 ml. If the volume to be injected is larger, injections of the maximum volume must be performed in different areas, or another route of administration must be evaluated.

To accomplish a subcutaneous injection, different techniques are recommended for nonanesthetized and the anesthetized rats. For nonanesthetized rats, it is preferable to make the injection 1 to 2 cm next to the spine in the dorsal area, with the help of an assistant who can hold the rat. If the rat is under the effect of general anesthetics, it is preferable to inject the solution in the abdominal area because the skin is thinner.

Despite the area of injection, the skin must be pressed with the thumb and the forefinger in order to create a fold, and the injection needle must be put in the fold between the fingers (Figure 11.2). Once the injection needle passes through the skin, it must be free to move between the thumb and the forefinger. While the solution is injected, the formation of a small bubble should be detected between the thumb and the forefinger.

For this route of injection, 15 × 0.5 or 13 × 0.4 needles are recommended.

Intraperitoneal Injection

An intraperitoneal injection is used when a volume higher than 1 ml is needed. The maximum volume for a 200-g rat is 5 ml. This route of administration has rapid absorption and can be used

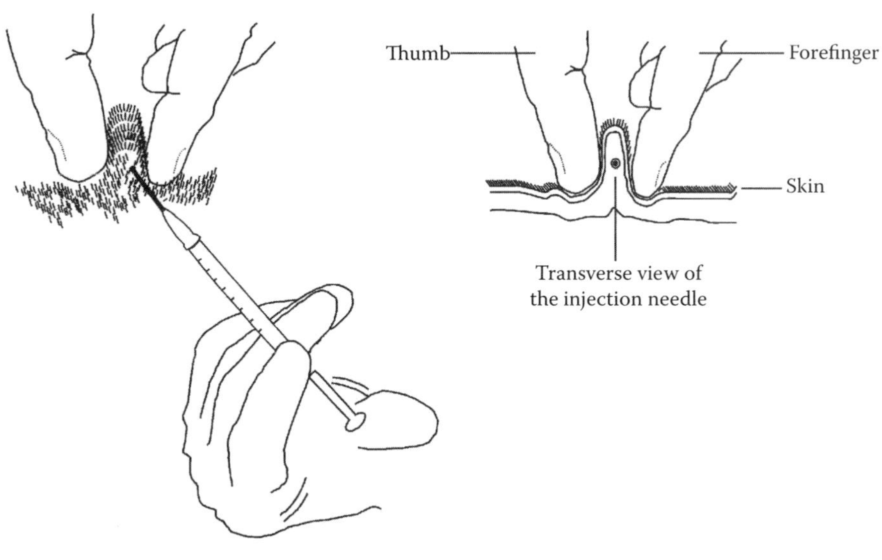

FIGURE 11.2 Pictured is the technique for the subcutaneous injection of a drug.

with different substances, such as urethane, glucose, amino acids, etc. An intraperitoneal injection of very active substances must be avoided, or very low doses used.

It can be done with anesthetized or nonanesthetized rats. In both situations, the injection is made with a 25 × 0.8 injection needle, which is introduced in a perpendicular direction to the skin in the abdominal area.

There are different ways of doing an intraperitoneal injection.

1. With two people: While an assistant holds the rat in a vertical position with both hands, the head and upper limbs of the rat must be immobilized with one hand, and tail and lower limbs with the other hand. The left hand of the person who is going to inject the solution presses the skin and makes a slight force on the skin to the left in order to produce tension in the skin. This maneuver makes it easy to prick the skin with the injection needle. With the syringe in the right hand, the needle is introduced in a perpendicular direction to the skin in the middle of the abdomen, halfway between the sternum and the pelvic area. The needle must be injected 1 to 2 cm into the animal, but must not be directed to the sides because of the risk of injection into the kidneys. If the needle is introduced more than 2 cm, there is risk of injecting the substance into the aorta or vein cava. The needle must be introduced in one movement in approximately 1 to 2 seconds. If done too quickly, there is a greater risk of injection into the intestine.
2. With only one person: This option is only recommend for skillful surgeons. One possibility is to hold the rat by the skin of the back with the left hand, producing a tension that prevents the rat from moving its limbs and tail, and then, with the right hand, use the maneuver with the needle and syringe as detailed above (Figure 11.3).

Another possibility is to put the rat on a coarse surface, so that the rat does not slip. Then with the left hand, the skin of the flank is caught and an ascendant movement is done to separate the right lower limb of the rat from the surface. This produces an immobilization of the rat. Finally, the injection needle is introduced into the abdominal area (Figure 11.4). In this maneuver, the rat is free to move. This method, however, is only recommended for very skillful persons. This technique produces less fear and stress than the other techniques described above.

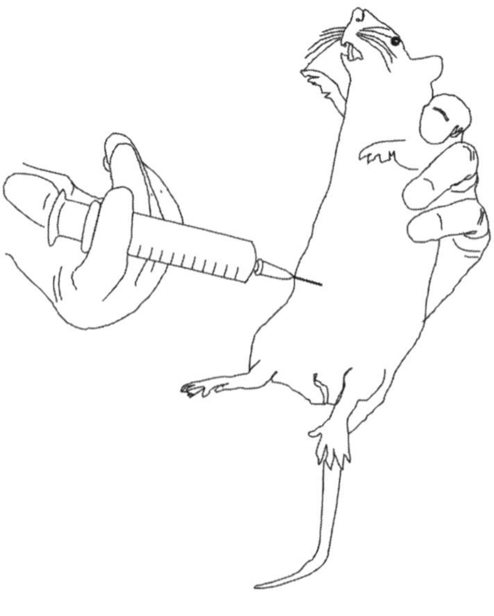

FIGURE 11.3 The technique for intraperitoneal injection by one person is shown.

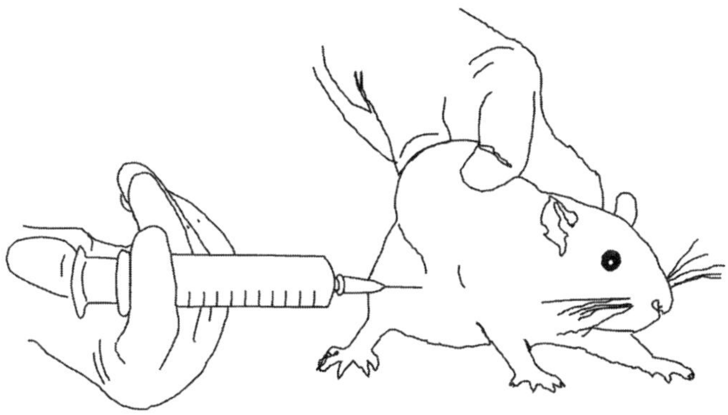

FIGURE 11.4 Shown is the technique for the intraperitoneal injection without the help of an assistant.

Injection in the Vein of the Tail

This technique is useful for the intravenous (i.v.) injection of a substance and allows injecting different volumes of solution. This procedure avoids the catheterization of an artery or vein, which requires incision and suture. One possibility is to inject directly using a syringe and it is advisable to administer approximately 0.1 to 1 ml. It is essential to take into consideration the rate of injection and this will depend on the properties of the substance. If this substance is not harmful to the rat, it could be injected at a higher rate. But in other instances where the substance is harmful, it should be injected slowly. Another possibility is to infuse a solution by cannula, e.g., hydration with physiological solution after a surgery when the rat is still under the effect of the anesthetic. The solution must be infused very slowly—by dripping, if it is a volume larger than 1 ml. It is important to control the osmolarity as well as the pH of the solution, and it is recommended that these variables be close to 300 mosmol/l and 7.4, respectively.

The injection must be done under the effect of anesthetics. Volatile anesthetics are recommended because the procedure takes only a few seconds. After anesthesia, the rat is put on a thermostatized rat board in dorsal decubitus position.

The tail is disinfected with ethylic alcohol in the area of injection and it is advisable to put on a band after the needle is removed to avoid blood loss.

The maneuver (explained in Figure 11.5) is for a right-handed person. The tail of the rat should be grabbed with the left hand, with the thumb above the tail and below the place that will be punc-

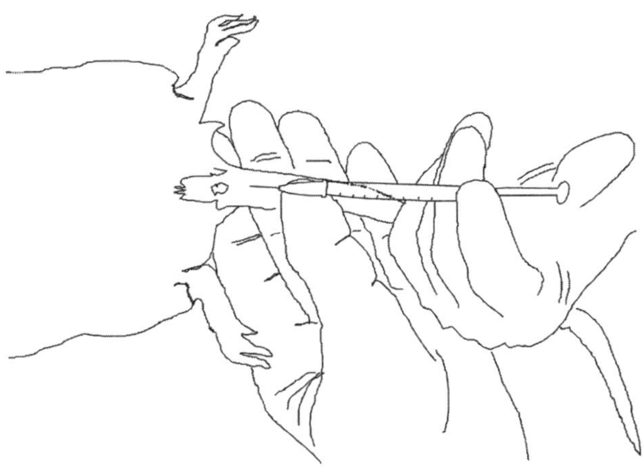

FIGURE 11.5 Injection in the vein of the tail is demonstrated.

tured, with the other fingers placed below the tail. The syringe is in the right hand. For a left-handed surgeon, just do the opposite.

The angle of injection should be approximately 30° to 45° and, when the needle reaches the vein, a drop of blood should appear where the needle entered. This would indicate that the vein was punctured and, thus, allows injecting the solution. In this situation, the injection is done without resistance. If the needle has missed the vein, a resistance to the injection is felt and swelling of the tail is observed.

For this injection, it is recommended that one use a 0.3 × 13 mm needle that is smaller than the ones described in the other techniques in order not to damage the vein. In addition, the needle should penetrate the skin 5 to 7 mm to reach the vein.

Intramuscular Injection

This route of administration allows injecting up to 0.25 ml for a rat of 200 g of body weight. It can be performed with a 13 × 0.4 or 13 × 0.3 injection needle and the best area is the muscles of the lower limb. The muscles of the lower limb are held between the thumb and the forefinger, and the needle is introduced into the muscles (Figure 11.6). The puncture should be perpendicular to the skin to properly introduce the solution. If the rat is not under the effect of anesthetics, the maneuver must be performed with the help of an assistant.

This route is used for the injection of anesthetics, antibiotics, and analgesics.

Intragastric Administration

The placing of the orogastric catheter can be performed individually or with the aid of an assistant to hold the animal. In both cases, it is necessary to hold the rat firmly so that it can keep its mouth open to allow the insertion of the catheter. It is necessary to be extremely careful so that the animal does not cut the catheter with its incisors. The catheter is introduced through the pharynx and must descend through the esophagus down to the stomach with no effort. In case the catheter takes the respiratory tract, a resistance to the passage is felt, and the catheter must be immediately removed. It is convenient to mark the catheter with the appropriate length from the mouth to the stomach. This mark is helpful to detect if the catheter is introduced in the trachea where the distance from the mouth is smaller. As an example, for a 250-g rat, a 10-cm catheter is appropriate.

The catheter must be made of polyethylene (BSN35) and be as flexible as possible in order not to harm the animal. It is normal for the rat to get stressed, as it cannot move, and it can cut the end

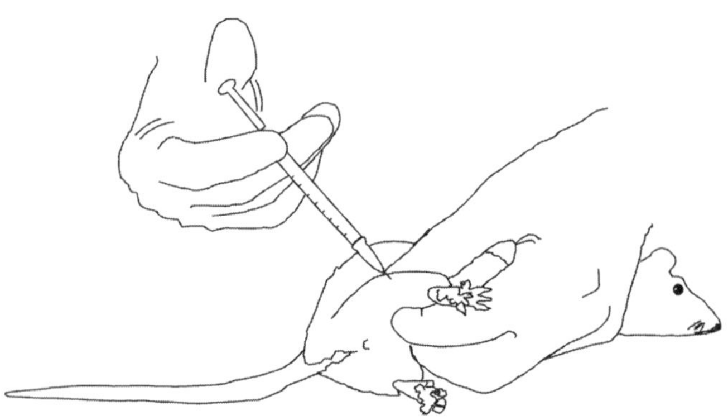

FIGURE 11.6 The technique for intramuscular injection with the rat under the effect of anesthetics is pictured.

of the catheter with its teeth. In this case, the catheter can still be used, but it must be controlled to make sure the area does not present sharp ends. If it does, they must be eliminated with the aid of a Bunsen burner.

Preferably, this technique should be performed with an animal that is conscious because the animal under anesthetics tends to regurgitate to the oral cavity and to produce asphyxia. Besides, a rat that is not under the effect of anesthetics makes the procedure of inserting the catheter easier, and it is easier to ascertain if it is inserted correctly.

The insertion of an orogastric catheter can be performed when it is necessary to hydrate the animal or when the administration of a unique dose of a substance is needed. However, there are some experiments that require a permanent orogastric catheter and the infusion of substances must be performed with the rat under anesthesia. In this case, the catheter must be placed in a conscious animal and then it must be anesthetized. Being extremely careful, the substance must be administered dissolved in a small volume. Leaning the rat board to avoid having the liquid ascend into the oral cavity is a good idea.

12 Samples

Verónica E. Di Loreto, Laura I. Pera, and Alfredo Rigalli

INTRODUCTION

Surgery in animals has two fundamental objectives: (1) to produce an experimental model or (2) to generate samples for a subsequent laboratory study. Urine and blood are the most common samples from experimental animals.

URINE

VESICAL CATHETERIZATION

Utility

Vesical catheterization is the elected method used to collect urine up to 6 hours in one or several fractions.

Materials

General anesthetic
Thermostatized rat board
Ring-handled preparation scissors
Ring-handled iris scissors
Adson forceps (2)
Artery forceps (2)
Needle holder
Orogastric polyethylene catheter BSN35
Polyethylene catheter BSN35 or similar (approximately 6 to 7 cm long)
Saline solution
Iodopovidone
Suture needle ½ – 5 to 10 mm.
Suture thread
Cotton balls

Procedure

Once the animal is under anesthesia, an incision is made on the skin (1.5 to 2 cm in length) and on the muscle at a suprapubic level (1 to 2 cm in length) (Figure 12.1).

The identification of the bladder is easier if it is full of urine. It is convenient to hydrate the animal before the surgery. Once the bladder is located, it is taken by the superior end with forceps, avoiding damaging the ureters, and with an iris scissor, a small cut is made in this end (Figure 12.2).

Still holding the bladder, a polyethylene catheter (BSN35) is inserted and the bladder is carefully tied to it, avoiding the binding of the ureters (Figure 12.3). It is convenient to perform an expansion in the end of the catheter that is inserted in the bladder, so that, once tied, it cannot slide out.

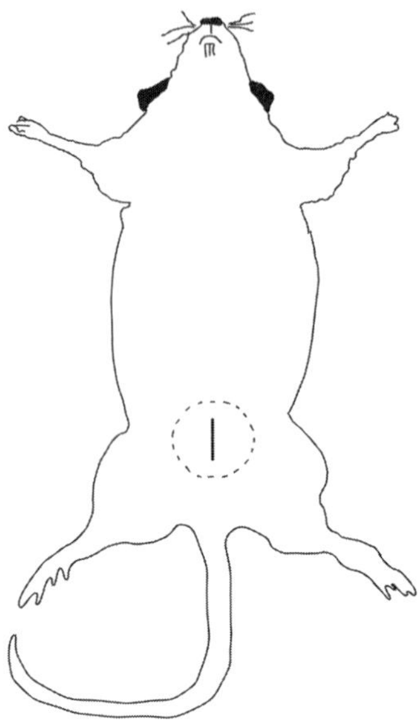

FIGURE 12.1 Area of incision for vesical catheterization.

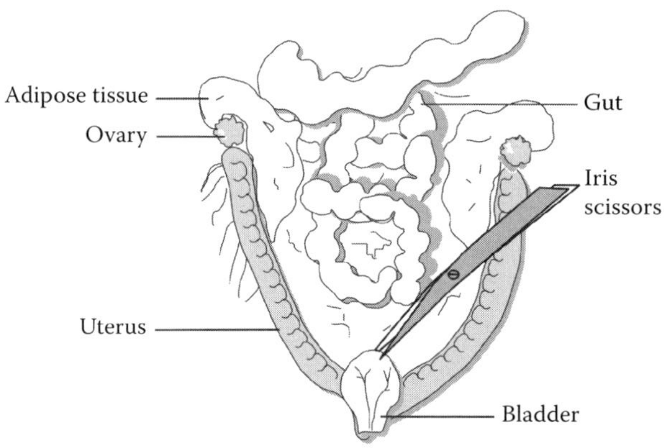

FIGURE 12.2 Incision in the upper end of the bladder is shown.

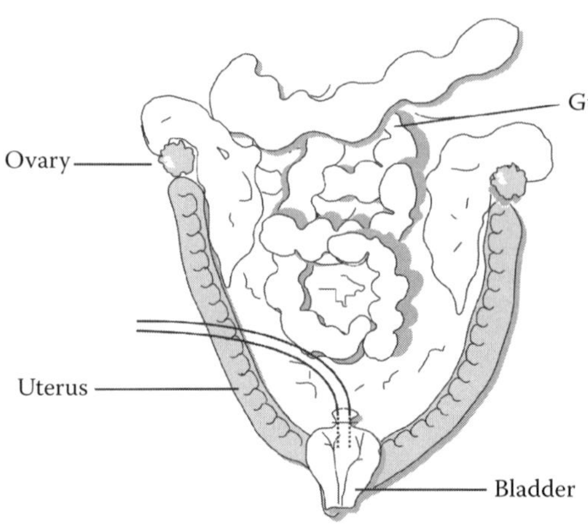

FIGURE 12.3 Pictured is the ligature of the catheter at the upper end of the bladder.

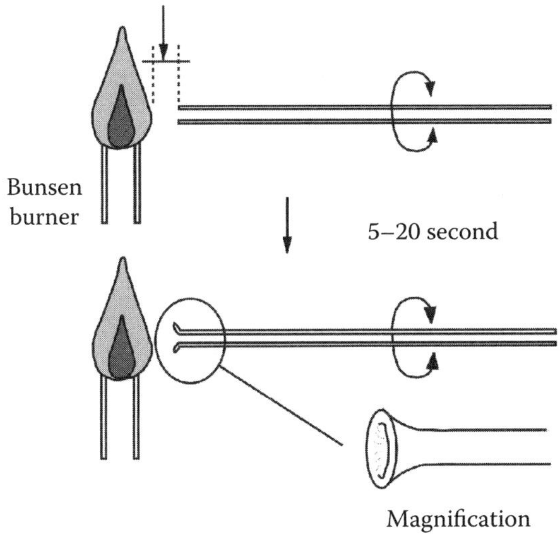

FIGURE 12.4 Preparation of the catheter for vesical catheterization is demonstrated.

To produce that expansion, the catheter must be cut perpendicularly and then the end placed close to the flame of a Bunsen burner (without touching it). Keeping the catheter between the forefinger and the thumb, it is spun (Figure 12.4). After a few seconds, a small ring, wider that the rest of the catheter, is formed. This ring is what prevents the catheter from sliding out of the bladder.

Special Considerations

One must be extremely careful at the moment of performing the incision on the bladder because this organ is extremely vascular. Therefore, the incision must be performed without cutting the blood vessels if possible or affecting the least amount of vessels possible to avoid hemorrhage. If it persists, urine will be contaminated with blood, invalidating the experiment. Also, blood in the urine can cause blood clots that can block the catheter. The heparinization of the catheter and/or its filling with saline solution with heparin and its subsequent elimination will avoid blood clot formation.

On experiments where the collection of urine is performed every 4 to 6 hours, it is necessary to rehydrate the animal through an orogastric catheter because the animal loses approximately 1.5 ml of urine/h 100 g body weight.

The loss of water through evaporation in the operating field and through the respiratory tract must be taken into account as well. It is recommended to rehydrate the animal at 2.5 ml/h 100 g body weight.

Experimental Results

An experiment is described below in which vesical catheterization was performed in rats with the aim of measuring urinary flow. Seven-week-old male rats (strain IIM/Fm substrain "m" with an average weight of 200 g) were used. After hydrating the animals through an orogastric catheter (5 ml water/100 g body weight), they were anesthetized with 120 mg urethane/100 g body weight. A catheter was placed in the bladder with the purpose of collecting urine during two periods of 30 minutes. The values of urinary flow of each rat are shown in Table 12.1.

As it can be observed, for the rats of approximately 200 g, the average urinary flows were 0.12 ± 0.04 ml/min for the first period and 0.11 ± 0.03 ml/min for the second period. It is clear that in relatively short periods of time, the flow remains steady.

TABLE 12.1
Urinary Flow

	Volume Minute (ml/min)		
Rat	**Period 1**	**Period 2**	**Body weight (g)**
1	0.14	0.13	197
2	0.12	0.15	205
3	0.17	0.08	187
4	0.12	0.13	168
5	0.07	0.09	169
6	0.14	0.10	203
7	0.09	0.12	203
8	0.05	0.06	178
9	0.13	0.09	175
10	0.15	0.14	214
11	0.11	0.13	198
Mean ± SD	0.12 ± 0.04	0.11 ± 0.03	

BLOOD

Blood can be withdrawn with the aim of analyzing it as a whole (e.g., hematocrit), studying its cells (works with red cells or leukocytes), or analyzing the plasma metabolites (proteins, glucose, triglycerides, etc). In any case, whole blood that can be subsequently separated in plasma and cells by centrifugation is obtained.

EXTRACTION FROM THE TAIL TIP

Utility

It is the elected method when the amount of blood needed is small (10 to 100 μl) or several extractions must be performed at different moments.

Materials

Ring-handled preparation scissors
Heparinized capillary glass tubes or centrifuge Eppendorf tubes with heparin
Modeling clay
Microcentrifuge
Ethlic alcohol
Cotton balls

Procedure

It is common to use this technique on a conscious animal, although it can also be performed with the animal under the effect of an anesthetic.

The animal is placed in the palm of the hand with its head pointing to the surgeon's elbow, leaving free the useable hand. The rat is held by the base of the tail with the hand in which the animal is resting. The tail tip must be cut off with scissors. This cut must be performed on the end of the

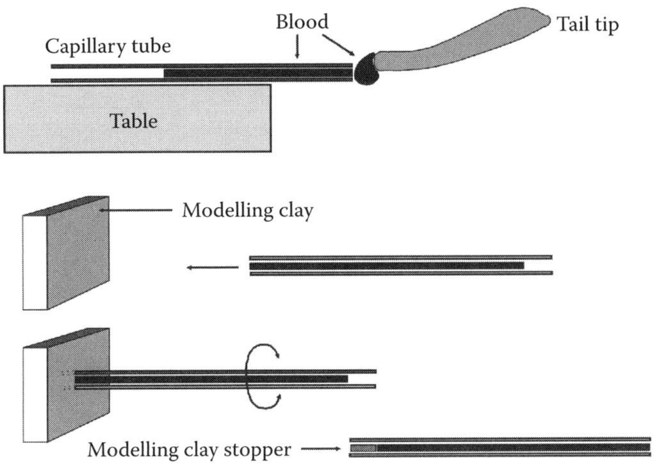

FIGURE 12.5 Shown is the handling of the animal and the capillary for the tail blood extraction.

tail so that only the skin is cut (equivalent to 1 to 2 mm of the tail). If the cut is well done, a drop of blood will form on the clipped end (Figure 12.5).

If this condition is reached, the blood is extracted using the free hand. With this hand, the base of the tail is held using the thumb, the forefinger, and the middle finger. They must slide toward the end of the tail pressing slightly. It can be observed that a drop of blood will form on the end of the tail. This drop can be collected in a small tube or in a capillary tube placed horizontally over the working table. If using capillary tubes, they can be filled just by touching the drop of blood at the tip of the tail. This must be repeated until obtaining the necessary amount of capillary tubes. If plasma is needed, once the tube is full, it is closed with modeling clay and centrifuged in a microcentrifuge for two minutes. Capillaries are cut at the level of the buffy coat (the joint cells–plasma) and the plasma is extracted with the help of a micropipette. This operation can be repeated at different times. If those periods are short, it is not necessary to cut the tail again. If it is not possible to obtain blood, the clot formed on the end of the tail must be removed by pressing with a piece of cotton and blood will flow without the necessity of cutting the tail once again.

Special Considerations

Modeling clay is kept in a metallic or plastic plain box of 10 cm^2 surface and 5 to 10 mm depth. With the box placed in a vertical position (see Figure 12.5), the glass capillary tube is plunged into the modeling clay, keeping it in a horizontal position. This last movement is repeated two to three times making the capillary tube spin between the fingers and pressing against the end of the box. In this way, the modeling clay is tightly introduced into the end of the tube and no blood is lost during centrifugation.

Extraction by Heart Puncture

Utility

This method is used when the amount of blood needed is more than 200 μl.

Materials

General anesthetic
Thermostatized rat board
1 to 10 ml syringe (according to the volume of blood)
Injection needle 25 × 0.8

Centrifuge tube
Centrifuge
Heparin (when plasma is needed) or any other anticoagulant
Iodopovidone
Cotton balls

Procedure

To perform the extraction, the animal must be anesthetized. It is impossible for experimental and ethical reasons to perform this kind of extraction without anesthetics.

Once the animal is under the effect of anesthetics, it is placed in a supine position with its legs completely extended over the rat board, creating a 45° angle together with the middle line of the body. In this position, it must be punctured in the intersection of the extension of the two imaginary lines that go through the axis of the forelegs and the middle line of the animal (Figure 12.6).

Once this spot is located, the forefinger is placed on it to search for the place where the heart beats are more intensely perceived. The needle is introduced in this spot, in a vertical direction, slowly and firmly, until a drop of blood appears on it. It is preferable to use a needle with a transparent plastic adaptor. The syringe is held firmly and the plunger must be pulled slowly. It is very important that, once the heart is located, the syringe be held firmly so that the needle remains in place because, if it moves, it will be hard to find it again and also it is possible to harm this organ.

It may happen that the needle penetrates to a deeper position, going through the heart. In this case, the needle must be removed slowly until the blood drop enters in it.

Special Considerations

For success in this method, the quality of the syringes must be taken into consideration. Using glass, high-quality syringes and those heparinized to lubricate the plunger make the movement easier. This allows the syringe to fill through blood pressure and assures the proper insertion of the needle in the heart cavity.

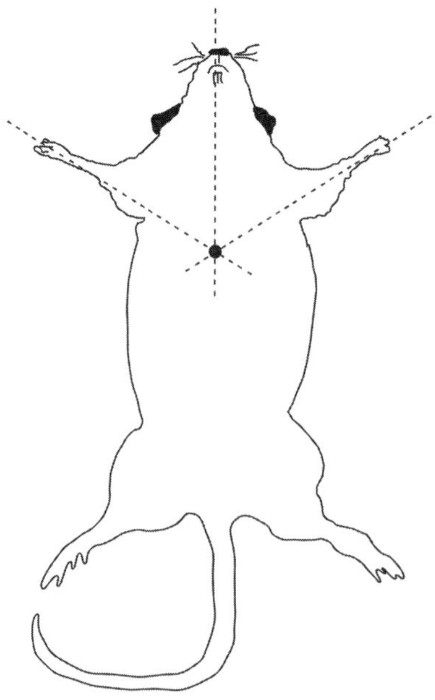

FIGURE 12.6 Location of the point for heart puncture is demonstrated.

Extraction through Artery Catheterization

Utility

Extraction through artery catheterization is applicable when large volumes of blood are needed (more than 200 μl) during a period of time in which more than one sample is needed.

Materials

- General anesthetic
- Thermostatized rat board
- Ring-handled preparation scissors
- Cotton pliers (2)
- Needle holder
- Spatula
- Catheter PC-40 or PC-50 with a beveled edge
- 1-ml syringe
- Heparin
- Iodopovidone
- Suture needle ½ – 5 to 10 mm.
- Suture thread
- Cotton balls

Procedure

(See arteriovenous catheterization in Chapter 13.) The femoral artery, because it is accessible, is the vessel chosen for this technique, but the carotid artery also can be used.

Section III

Catheterization and Cannulation

13 Catheterization of the Femoral Artery and Vein

Verónica E. Di Loreto and Alfredo Rigalli

CATHETERIZATION OF THE FEMORAL ARTERY AND VEIN

UTILITY

Catheterization of the femoral vein is used to perform the injection or a continuous infusion of a substance. On the other hand, catheterization of the femoral artery is used to extract several samples of blood in volumes higher than 100 μl or to measure the artery pressure.

Materials

General anesthetics
Thermostatized rat board
Ring-handled preparation scissors
Cotton pliers (2)
Needle holder
Spatula
PC-40 catheters with beveled end for artery
PC-40 or PC-50 catheter with beveled end for vein
1-ml syringe
Injection needle 15 × 0.5
Saline solution
Heparin
Iodopovidone
Suture needle ½ – 5 to 10 mm.
Suture thread
Cotton balls

The preparation of the catheters is shown in Figure 13.1.

The catheter must be cut with the ring-handled preparation scissors in a 45° angle and the beveled end of a 15 × 0.5 injection needle must be cut with the aid of a saw. The blunt end of the needle must be polished with sandpaper or a whetstone. In this way, the damage to the catheter is avoided.

PROCEDURE

The rat is anesthetized, placed in supine position, and an incision is made in the inguinal area (Figure 13.2). Then, with cotton pliers, the connective tissue is separated until the vascular nerve package formed by the femoral vein, the femoral artery, and the nerve can be seen. The vein has a dark red color, the artery is clearer and brighter than the vein and the beats can be observed. The nerve has a whitish tone.

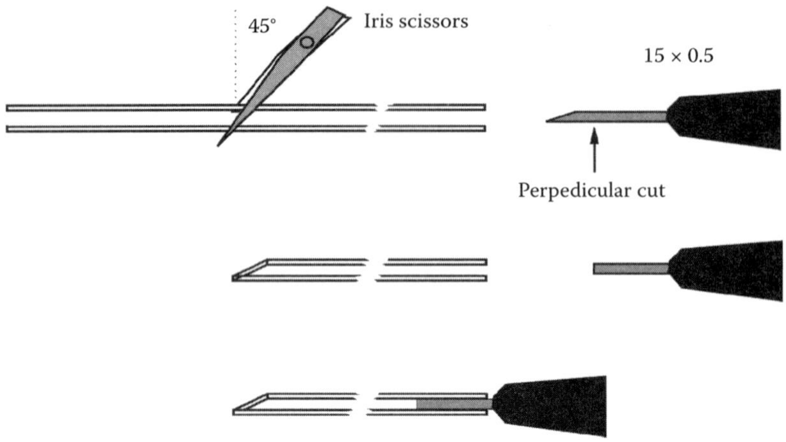

FIGURE 13.1 Assembly of beveled catheters with the needle.

The following step is to separate these three components. The separation can be performed by sliding a spatula under the vein, artery, and nerve, with the aid of a cotton plier. The three components are carefully separated, avoiding damage (e.g., pricking the artery or the vein). The separation can also be performed directly with a forceps without the need of a spatula.

Once these elements are separated, it is convenient to slide a thread under the artery as well as under the vein in order to keep them separated and to use it to hold the catheter. Once the artery and vein are separated, the cannulation is performed. Two pieces of thread and a spatula are slid under the artery (Figure 13.3a).

The distal end is bound and also pulled to cause a slight stretching of the artery (Figure 13.3b). The spatula is rotated 30 degrees and an orifice is made with a 15 × 0.5 injection needle (Figure 13.3c). When removing the needle, the spatula is turned in the opposite direction to block the blood flow. If the orifice is properly performed, a blood drop will form (Figure 13.3d).

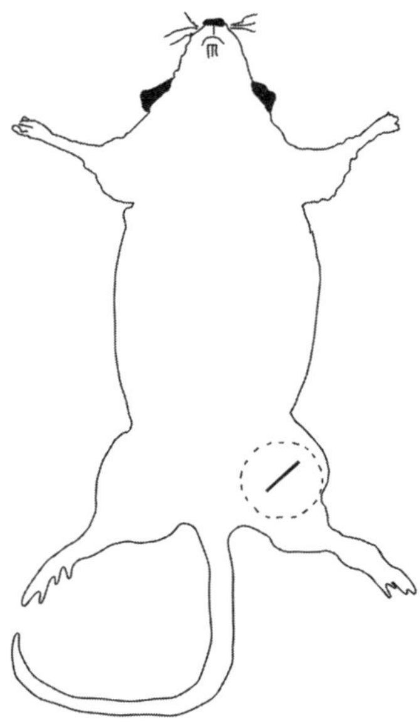

FIGURE 13.2 Area where the incision is made when performing the femoral catheterization.

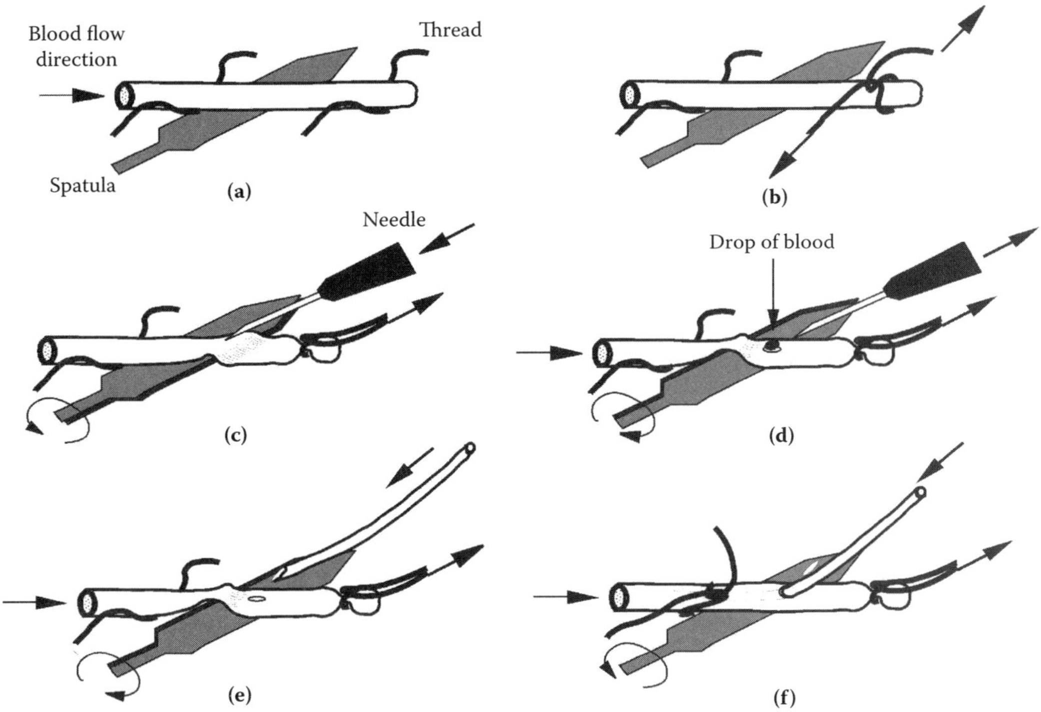

FIGURE 13.3a–f The procedures for artery catheterization.

Keeping the spatula in the same position, the catheter is inserted and it returns to the previous position (Figure 13.3e). Once cannulated, the catheter is bound to the vessel to keep from sliding out (Figure 13.3f).

The control of the proper catheterization is easy to determine. In the case of the artery, once cannulated, the blood goes out of the catheter because of its pressure. That is why it is necessary to place a heparinized syringe on the other end to stem blood flow, and it will also be used to extract samples. In the case of the vein, the control of the proper placing of the catheter can be made in two ways: (1) extracting with a syringe to make the blood flow (determining that there are no air bubbles because that would mean the vein might have been damaged) or, (2) if the substances administration will be performed with an infusion pump, it can be determined whether the catheterization was correctly performed or not, changing the flow direction.

Special Considerations

The incision made on the skin of the rat must not be sutured because, if complications arise (the catheter slides out, for example), there is a faster access to the vessels. But, as there is no suture, a piece of cotton with saline solution must be placed in the open area to avoid the evaporation of body fluids.

Care needs to be taken to avoid the collapse of the vessels, especially the vein, because this can make the cannulation process difficult. That is why, during the handling of the vessels, excessive stretching must be avoided. When sliding the thread under the vessels, it must not scrape them because this can cause the entangling of the vessel and can inhibit the subsequent extraction of samples.

If a large number of samples are to be extracted, it may be convenient to heparinize the animal. In this case, 500 U heparin/kg of body weight is used, which equals 5 mg heparin/kg of body weight.

A syringe is filled with the solution and heparin is injected immediately after cannulation through the catheter, in the artery or vein. In this way, neither the animal nor the material need be reheparinized. The reinjection of red blood cells after the centrifugation and collection of plasma samples is recommended in cases where a large volume of blood must be extracted during the surgery (more than 2 ml). To perform this procedure, the blood must be extracted with heparin as indicated above. Centrifuge and separate the plasma, resuspend the red blood cells in the same volume of isotonic saline solution, and reinject this solution. With this technique, the survival of the animal is increased, especially in surgeries in which blood volume is severely reduced by hemorrhage or samples extraction.

14 Cardiac Catheterization

Cristina Lorenzo Carrión, Laura Krieger, Manuel Rodríguez, and Martín Donato

CARDIAC CATHETERIZATION

UTILITY

In 1844, Claude Bernard inserted a mercury thermometer into the carotid artery of a horse and advanced it through the aortic valve to the left ventricular chamber in order to measure blood temperature. He adapted this experiment over the next 40 years for measuring intracardiac pressures in a variety of animals. It is because of his work that the use of catheters became the standard method for physiologists in the study of cardiovascular hemodynamics. At this time, cardiac catheterization continues to evolve and expand, including advances in the fields of material science and miniaturization. Thus, it is an important technique, useful to evaluate ventricular function through the analysis of arterial and ventricular pressure curves.

With the catheter placed into the carotid artery, it is possible to obtain recordings of the systolic blood pressure (SBP) and the diastolic blood pressure (DBP). With this, values of the mean arterial pressure (MAP) is calculated:

$$MAP = \frac{1}{3}(SBP - DBP) + DBP \tag{14.1}$$

With the catheter inside the left ventricular chamber, it is possible to obtain intraventricular pressure recordings and first derivatives (dP/dt). Considering the left ventricular systolic pressure (LVSP) and the left ventricular end diastolic pressure (LVEDP), it is possible to calculate the developed pressure (LVDP): LVDP = LVSP – LVEDP (Figure 14.1). These hemodynamic indexes allow a detailed evaluation of the left ventricular function, either their systolic or diastolic components.

Materials

General anesthetics
Local anesthetic
Thermostatized rat board
Iridectomy scissors
Microforceps (2)
Needle holder
Weitlaner–Locktite retractor
Alm retractor
Clamp
Pressure transducer (e.g., from Becton, Dickinson and Company, Franklin Lakes, New Jersey, U.S.A.)
Register and acquisition data equipment

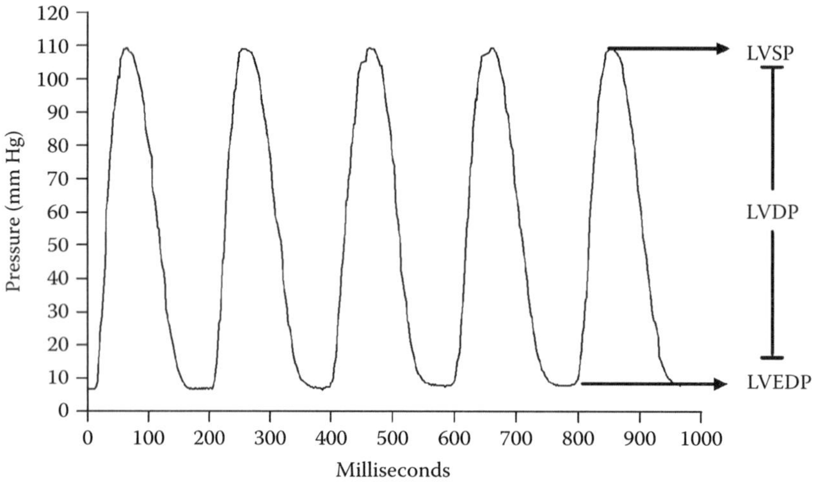

FIGURE 14.1 Left ventricular pressure is shown.

Epidural anesthetic catheter 16 to 20 G (depending on the diameter of the carotid artery) (e.g., Perifix®, B. Braun Medical Inc., Bethlehem, Pennsylvania, U.S.A.)
Saline solution
Heparin
Iodopovidone
Suture thread
Gauze (10 × 10 cm)

Procedure

The animal is anesthetized with an intraperitoneal injection mixture of ketamine and xylazine; it is placed in dorsal decubitus and the ventral face of the neck is shaved. The medial region of the neck is infiltrated with a subcutaneous injection of lidocaine and a medial incision is made longitudinally in the ventral face of the neck. The different anatomic planes under the skin are divulsed (parted) until right sternomastoid muscle is exposed. This muscle is laterally tractioned to expose the right carotid artery with an Alm retractor. The right common carotid artery is isolated from the rest of the vascular components with an approximately 2-cm-long extension. Two linen threads are passed below the artery: one in the proximal end and another in the distal end (cephalic). The carotid artery is tied in the distal end and a clamp is placed in the proximal end. The artery is partially sectioned in a cross-sectional way using the iridectomy scissors in the medial sector between the clamp and the ligature. The catheter is connected to the pressure transducer and it is heparinized to prevent the formation of clots. This maneuver is made with extreme care, avoiding the formation of bubbles inside the catheter and pressure transducer. Next, the catheter is introduced into the artery lumen and slid forward until it is near the clamp. The clamp is gently removed, and the catheter is slowly slid forward. It is important to take care that the passing of the catheter is not obstructed and that the artery is not torn. The catheter is gently slid forward while observing the pressure registry because the morphology of the curves and the pressure values demonstrate the position of the catheter (Figure 14.2).

If there is difficulty getting the catheter into the ventricular chamber, it is recommended that slow in-and-out movements are made and the catheter is rotated in both directions. It is important to keep in mind that the catheter should be introduced at the moment of systole, when the valve is open, to avoid the tearing of the valves because that would cause aortic insufficiency and alteration in the hemodynamic values. Once across the valvular ring, the catheter is slowly slid forward, taking care

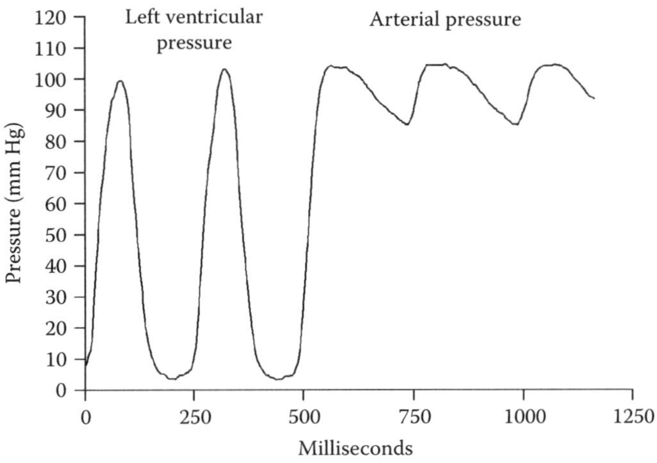

FIGURE 14.2 Typical recording of left ventricular and arterial pressures is shown. The catheter was moved from the ventricle to the aorta.

not to perforate the left ventricular wall. Sometimes, it is possible to feel a light vibration when the aortic leaflets strike the catheter. The entrance of the catheter into the ventricle is demonstrated by the observation of the typical left intraventricular pressure curve (Figure 14.3). Once within the left ventricular chamber, the catheter must be fixed to the proximal end of the artery by tightening the proximal linen thread.

Special Considerations

The size of the artery should be slightly larger than the diameter of the catheter. The artery must not be completely sectioned or should not have a cut that withdraws the artery when it is pulled to place the catheter. The clamp must be removed with extreme care because it could be adhered to the surrounding tissues and, if removed abruptly, it could damage the vessel wall.

When advancing the catheter into the vessel, care must be taken not to dissect or produce deendothelialization. This can be noticed by the absence of pressure recording. When trying to enter the left ventricle through the aortic valve, an excess of strength could perforate the aortic wall. This would also be noticed by the absence of pressure recording. When the catheter is in the left ventricular chamber, the ventricular pressure curve must have a typical morphology. The shape of the curve can be modified by the contact of the catheter with the ventricular wall. In that case, it is recommended to move the catheter softly until the ideal condition is obtained.

Experimental Results

As mentioned, this technique allows measured changes in hemodynamic variables, such as blood pressure and heart rate, and ventricular function. Below is an example of the behavior of arterial blood pressure and contractile state during hemorrhagic shock.

In Wistar rats, after 15 minutes of stabilization, we induced an acute hemorrhagic shock by withdrawing blood (1.4 ml/100 g body weight) during 2 minutes.

Table 14.1 shows the systolic blood pressure (SBP), diastolic blood pressure (DBP), mean arterial pressure (MAP), and heart rate (HR) during the hypovolemic shock. The hemorrhage induced an acute and significant decrease in arterial blood pressure during the first 2 minutes following the bleeding ($p < 0.05$ versus basal values). The arterial blood pressure values were

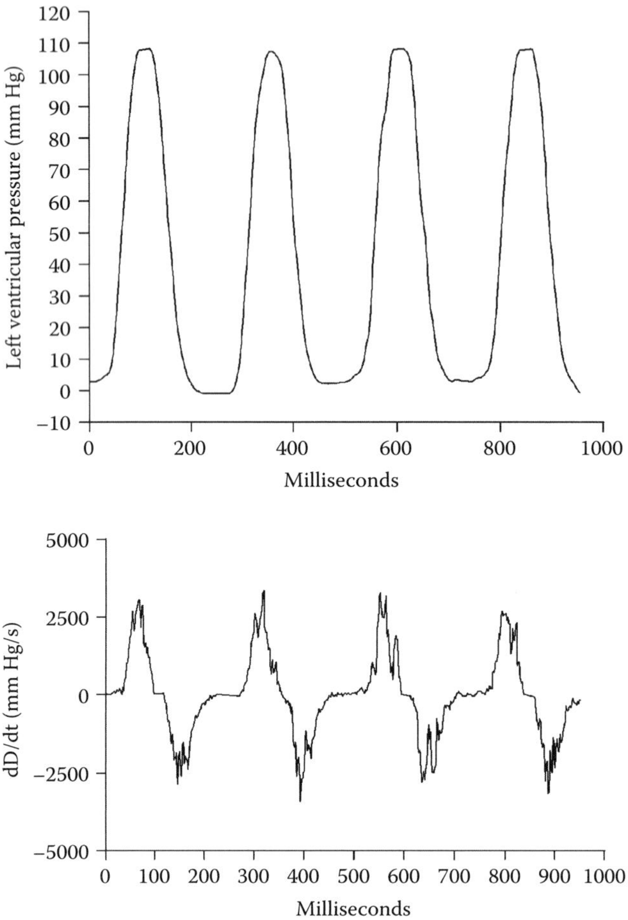

FIGURE 14.3 Recording of left ventricular pressure (upper panel) and its first derivative (lower panel) is pictured. The shape of the curve is similar to that observed in an isovolumic heart; the reason for this is the high heart rate of the rat.

recovered at 120 minutes after the hemorrhage. It is important to mention that the DBP decreased under physiological myocardial perfusion pressure values reaching 28.8 ± 2.5 mmHg at 2 minutes after bleeding. Thus, this finding could indicate the presence of a certain grade of subendocardial ischemia.

Table 14.2 shows the behavior of left ventricular function during bleeding and recovery. The hemorrhage decreased left ventricular developed pressure (LVDP) during bleeding. At the recovery, the LVDP reached similar basal values. On the other hand, for left ventricular end diastolic pressure (LVEDP), no significant changes were observed during bleeding or recovery in this parameter. Although, LVEDP increased slightly at 60 seconds of the bleeding, this parameter remained within the physiological values.

TABLE 14.1
Hemodynamic Variables during Hemorrhagic Shock and Recovery

Parameter	Basal	1 min	2 min	60 min	120 min
SBP (mmHg)	99.7 ± 7.5	67.3 ± 3.1	47.1 ± 12.1*	89.4 ± 10.2	103.3 ± 12.8
DBP (mmHg)	60.5 ± 6.2	34.4 ± 4.8*	28.8 ± 5.5*	40.2 ± 10.5	43.8 ± 9.4
PAM	76.6 ± 5.1	45.5 ± 4.7*	31.3±6.1*	56.6 ± 10.3	63.6 ± 9.3
HR beats/min	436.8 ± 10.1	392.4 ± 27.9	360.1 ± 84.3	422.2 ± 58.8	441.1 ± 7.1

Note: SBP: systolic blood pressure, DBP: diastolic blood pressure, MAP: mean arterial pressure, HR: heart rate.
* Indicates significant differences with respect to basal, p <0.05. Results are expressed as mean ± SEM.

TABLE 14.2
Ventricular Function

Parameter	Basal	1 min	2 min	60 min	120 min
LVDP (mmHg)	140.1 ± 1.4	130.7 ± 7.7	107.2 ± 2.14*	136.4 ± 4.5	125.7 ± 8.7
LVEDP (mmHg)	2.85 ± 0.0	3.35 ± 1.7	1.93 ± 0.8	1.57 ± 10.4	1.93 ± 0.8

Note: LVDP = left ventricular developed pressure; LVEDP = left ventricular end diastolic pressure.* Indicates significant differences with respect to basal; p <0.05. Results are expressed as mean±SEM.

15 Cannulation of the Thoracic Duct

Gabriel A. Inchauspe and Alfredo Rigalli

CANNULATION OF THE THORACIC DUCT

Utility

The thoracic duct collects lymph from the lower half of the body and conveys it to the blood stream. Lipids absorbed in the intestine are carried to the blood through this duct. Cannulation or the puncture of the thoracic duct is useful in studies about gastrointestinal digestion and intestinal absorption of lipid-soluble substances. Even though the permanent cannulation of the thoracic duct in animals has been executed, it is not easy to perform in the rat. Instead, the puncture of thoracic duct is preferable.

Materials

General anesthetic
Thermostatized rat board
Ring-handled preparation scissors
Ring-handled iris scissors
Cotton pliers (2)
Adson forceps
Artery forceps (6)
Rat-tooth forceps
Capillary tubes for hematocrit (according to the number of samples)
Modeling clay
Orogastric catheter
5-ml syringe
Injection needle 15 × 0.5
Saline solution
Iodopovidone
Cream from cow milk
Cotton
Cotton buds (10)
Gauze

The surgery materials mentioned above are needed for the puncture of the thoracic duct. Additional material must be considered if the technique is combined with another, such as cannulation of an artery, orogastric administration, etc.

Procedure

Four hours before surgery, 2 ml/100 g body weight of cream from cow milk is administered to rats by an orogastric tube. The administration makes it easier to observe the thoracic duct and also will increase the lymph flow. Following the ingestion of the cream, animals are anesthetized and a 3 to 4 cm median laparotomy is performed in the epigastric area. Artery forceps are used to open the area of surgery, and the intestine is moved to the left until the vein cava is observed. The latter maneuver must be done with cotton buds or cotton pliers. The vein cava, upper mesenteric artery, and celiac artery are easily observed. The thoracic duct and the upper mesenteric artery are parallel and close together. The former is located in the cephalic direction. The duct is easily distinguished because of the white color and position. To withdraw samples, a small hole is made in the duct with a 15 × 0.5 injection needle. The lymph flows easily and can be collected with a hematocrit capillary tube in the horizontal position. Once the sample is obtained, both ends of the capillary must be closed with modeling clay (see Chapter 12).

It is convenient to calculate the ratio volume/length of the hematocrit capillaries to be used. This can be done with the aid of a Hamilton syringe or other device for the measurement of small volumes. This ratio also can be calculated by the subtraction of the weight of the capillary from the weight of the capillary full of water. In the regular capillaries, 1-cm in length contains 10 µl of aqueous solutions.

Between the extractions of samples, the abdomen must be temporarily closed with the aid of 2 or 4 artery forceps. When obtaining a sample, it is convenient to eliminate the lymph fluid that could be lost through the thoracic duct. In the described method, it is possible to obtain samples for approximately 2 hours. After that time, the lymph flow decreased dramatically and the visualization of the duct is not clear.

Experimental Results

An increase in the plasma triglycerides is detected in rats after giving intravenously 2 µg of human recombinant growth hormone (hrGH).

The aim of this work was to study the role of the intestine in the increase of plasma levels of triglycerides after the administration of hrGH. Two hundred-gram female rats (IIM/Fm strain "m") were laparotomized as stated above. Lymph samples were extracted every 10 minutes for 1 hour. The femoral artery was cannulated at 10 minutes as stated in Chapter 13, and then, through this catheter, 0.5 ml of isotonic saline solution containing 2 µg de hrGH was injected. Control rats were subjected to the same maneuvers and received 0.5 ml of saline solution without hrGH.

The volume of lymph was measured by the ratio volume–length and a 1/10 dilution with saline solution was used for the measurement of triglycerides concentration. The change of triglyceride levels in lymph is displayed both for hrGH-treated and control animals (Figure 15.1).

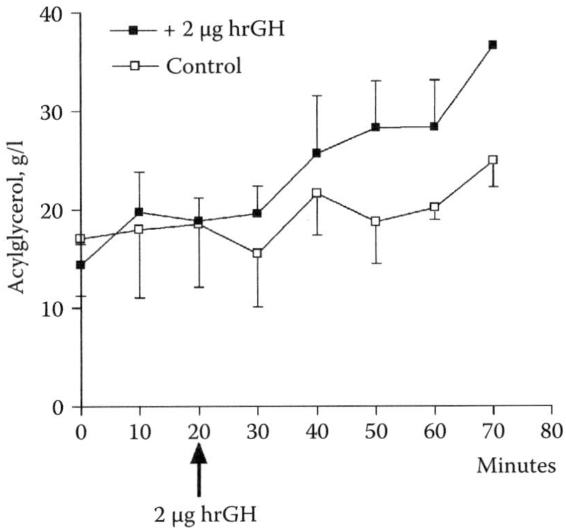

FIG. 15.1 Plasma levels of triglycerides in lymph after an intravenous dose of human recombinant growth hormone are shown.

16 Tracheostomy

Laura I. Pera and Alfredo Rigalli

TRACHEOSTOMY

Utility

The tracheostomy is applicable to numerous purposes. One of them could be in helping the respiratory function when a respiratory obstruction occurs by mucus accumulation as a consequence of the anesthetic. Another application may be the administration of gaseous or volatile substances like hydrogen, for example, with the aim of measuring the blood flow in the tissues.

Materials

- General anesthetic
- Thermostatized rat board
- Ring-handled preparation scissors
- Ring-handled iris scissors
- Cotton pliers (2)
- Rat-tooth forceps (1)
- Needle holder
- 3 to 4 cm in length beveled-end catheter with a diameter appropriate to the size of the animal (for a rat of 200 g, a catheter BSN35 will be suitable)
- Iodopovidone
- Suture needle ½ – 5 to 10 mm.
- Suture thread

Procedure

The rat is held in a supine position and a 3-cm longitudinal incision is made at the neck level (Figure 16.1).

The skin is cut with the curved ring-handled preparation scissors and then the muscular tissue planes are separated with cotton pliers or iris scissors until reaching the trachea. Cotton pliers are placed under the trachea that is slightly lowered so that a thread can be slid under it with the aid of the pliers. The trachea is lifted by taking the ends of the thread (Figure 16.2a), pulling in the opposite direction to the catheter inlet, and making a small orifice on it just between the two cartilaginous rings.

Then the catheter is inserted so that the beveled end heads toward the lungs, leaving the other end free (Figure 16.2b). For this process, the thread should be pulled through the trachea in the opposite direction to the inlet of the catheter, and a knot made tying both of them together (Figure 16.2c and Figure 16.2d), so that the catheter remains fixed in the correct position.

Precautions need to be taken to avoid also tying the esophagus, which is located under the trachea. If this happens, some difficulties could arise in the experiment, particularly if another substance needs to be administered through orogastric catheterization.

FIGURE 16.1 The area of incision for tracheostomy.

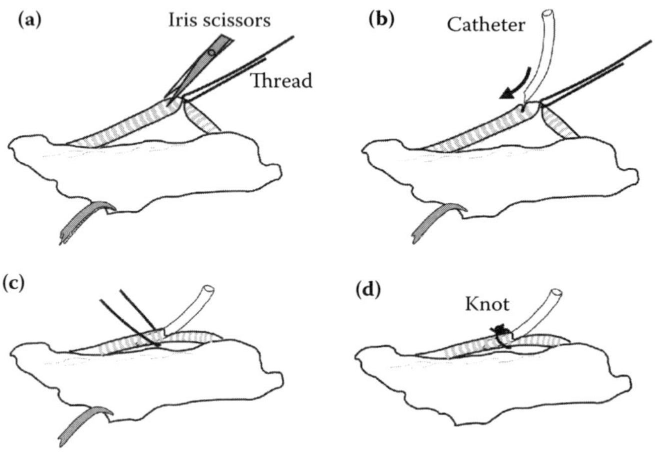

FIGURE 16.2 The tracheostomy technique.

Section IV

Gastrointestinal Tract

17 *In Situ* Isolation of the Stomach

Alfredo Rigalli

IN SITU ISOLATION OF THE STOMACH

UTILITY

This technique is useful in studying the gastric absorption of substances, their interaction with other components of the gastric content, and the mechanism involved in their degradation in the gastric lumen or the pharmacological actions on gastric mucous. This method shares several characteristics with the *in situ* isolation of the intestinal loop,[1] which is described below in this section. In the 1990s, it represented an invaluable technique to clarify the pharmacokinetics of sodium monofluorophosphate and sodium fluoride.[2-4]

The study of absorption in the *in situ* isolated stomach is complemented by the implementation of mathematical models developed for each purpose.[5] If these models are developed on firm foundations, a large amount of information about the processes occurring in the stomach lumen will be obtained.

As in the isolated intestine, the *in situ* isolated stomach is not separated from the animal, which remains under anesthetics during the experiment, maintaining the irrigation and innervation of the organ. At the same time, due to the anatomical characteristics of the organ, it has an advantage over the intestine, as it can be studied in its entirety. The technique allows the obtaining of serialized samples of stomach content as well as modifying the composition of the solution.

Materials

General anesthetic
Thermostatized rat board
Ring-handled preparation scissors
Ring-handled iris scissors
Cotton pliers (2)
Adson forceps
Artery forceps (3)
Rat-tooth forceps
Needle holder
BSN35 polyethylene catheter for orogastric catheterization (15 cm in length)
K31 polyethylene catheter for stomach cannulation (2 to 3 cm in length)
1-ml syringe (2)
10-ml syringe (2)
Saline solution
Iodopovidone
Filling solution of gastric lumen
Suture needle ½ – 5 to 10 mm
Suture thread
Cotton buds
Gauze

Procedure

The animal is anesthetized with a general anesthetic. Once the animal is under the effects of the anesthetic, it is shaved and the antisepsis of the incision area of the abdomen is performed while waiting for the analgesic and muscle relaxant to take effect. Then, a median laparotomy is performed of approximately 4 to 5 cm, depending on the animal's weight. The stomach is located in the upper left area of the abdomen, below the left lobe of the liver and above the small intestine. It can be partially covered by the spleen. In case it is not visible, it is recommended to locate the intestine and follow it until the pylorus is found; this sphincter is clearly distinguishable because it has a ring with a different color from the stomach and the intestine. The stomach is large, has a greater diameter, and has a grayish color with stretch marks, while the intestine is tubular and has a pink or reddish color. Once the pylorus is identified, a double thread is passed under the first part of the duodenum with cotton pliers and a knot is made, without tightening it. A 45°-angled flap incision is made on the duodenum, next to the pylorus, through which a catheter is introduced with a syringe on the other end to extract the samples (Figure 17.1).

Another option consists of binding the knot in the pylorus and extracting the samples through an orogastric catheter that can be kept in the animal during the surgery. This option is not as convenient because the catheter has a greater dead volume. As a consequence, this dead volume does not allow the liquids to be in contact with the gastric mucosa, and the samples obtained will not truly represent the processes that are occurring in the lumen of the organ. This inconvenience can be minimized with the catheter in the pylorus because it can be shorter, reducing the dead volume. For a 15-cm-long orogastric catheter with a radius of 0.1 cm, the volume of the content is approximately 0.5 ml. Taking this into consideration, 2 and 3 ml can be placed in the gastric lumen; this represents approximately 20% of the volume that is being studied. Using a catheter of 2 to 3 cm in the pylorus, the dead volume will not exceed 5%. Furthermore, being shorter, it is more difficult for it to become blocked and the catheter can be used to mix the content. Normally, part of the content is extracted with the syringe and then reinjected into the stomach. The catheter for the pyloros should be long enough to allow the correct binding and clamping with an artery forceps the moment the syringe is removed. It is convenient that the catheter has lateral holes and a nonsliding ring. This ring can be

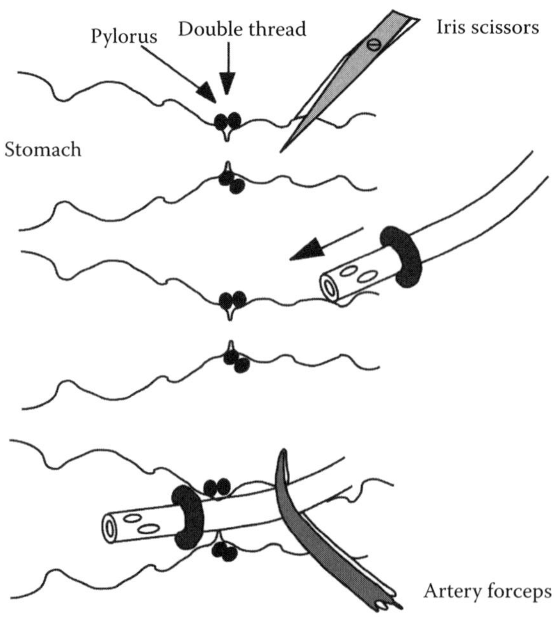

FIGURE 17.1 Pictured is the insertion of the catheter into the pylorus area.

easily made by winding firmly tied surgery thread onto the polyethylene catheter, or it can be done as explained for the vesical catheterization in Chapter 12.

The solution to be studied can be introduced through the catheter and, if necessary, the samples can be extracted through it as well. To avoid losing the content of the gastric lumen, an artery forceps should be used to clamp the catheter, so that the rest of the fluids are not lost through the catheter when removing the syringe with the sample (see Figure 17.1).

Precaution: The threads should be firmly tied, but should not place any excessive pressure on the tissues, as this causes bleeding and, in extreme cases, blood clots in the gastric lumen, preventing obtainment of reliable data. It is also convenient that, when instilling the solution, the stomach walls are not extremely tense, as this may cause difficulty in blood circulation.

The amount of samples obtained will depend on the volume of each sample as well as on the filling volume. Approximately 2 to 3 ml of solution can be used for the stomach of a rat of 200 to 250 g body weight. In the case of nonfasted animals, it is possible to fill the lumen with solution, extracting this solution and eliminating it, and then filling the lumen with the definitive solution. In this way, the rest of any food is eliminated, as any remaining food could make the determinations as well as the extraction of samples difficult.

If substances must be added throughout the experiment, it is convenient not to remove the orogastric catheter. This also helps in avoiding the reflux of the gastric content. It is unnecessary to make a binding at the level of the cardia when the volume of the solution added is small. If the volume is large and produces a great stomach distension, it is convenient to make a binding at the cardia level to avoid losing the solution.

A valid alternative would be to make a cut such as the one made in performing a tracheostomy, identifying the trachea, to locate the esophagus from its back side. With the aid of cotton pliers, a thread can be passed through the esophagus and a knot can be made in that area. The technique used will depend on the skills of the surgeon and the experimental requirements.

When it is necessary to know the concentration of a substance in the gastric lumen, the changes of volume that can occur must be taken into account, due to secretions as well as to the absorption of water produced in this area. The changes of volume can be controlled, as explained in Chapter 18.

Experimental Results

The aim in this experiment was to investigate the process of gastric absorption of sodium fluoride (NaF) and sodium monofluorophosphate (MFP). The rats used were female 190 to 230 g body weight, adults (IIM/Fm) subline "m."[6] Rats were intraperitoneally anesthetized with urethane (120 mg/100 g of body weight). During the experiment, the room temperature was kept between 21 to 22°C. The rats were placed on a thermostatic board to maintain their body temperature.

A cut was performed in the pylorus, through which a catheter was placed as described above. An orogastric catheter was introduced, through which a solution of MFP 52 mM and NaF 13 mM was instilled (Table 17.1). The concentration of fluoride was measured by direct potentiometry with an Orion electrode 94-04. The concentration of MFP was determined using a technique developed in the laboratory.[2] The results are shown in Figure 17.2.

Starting from these data, applying a mathematical model[5] and a combination of experiments, the value of the rate constant of MFP absorption in the stomach was obtained. Absorption followed a first-order kinetic and the value of the rate constant resulted in $0.0082 \pm 0.0010\ min^{-1}$.[2]

TABLE 17.1
Concentrations of MFP and F in the Gastric Lumen after Placing 2 ml of MFP Solution

Time (min)	[MFP] mM	[F] mM
0	52 ± 5	13 ± 2
15	48 ± 4	13 ± 5
30	37 ± 7	10 ± 2
45	37 ± 4	9 ± 1
60	35 ± 6	8 ± 2
75	30 ± 1	8 ± 1
90	23 ± 2	7 ± 1

Note: The results are expressed as the mean ± SD.

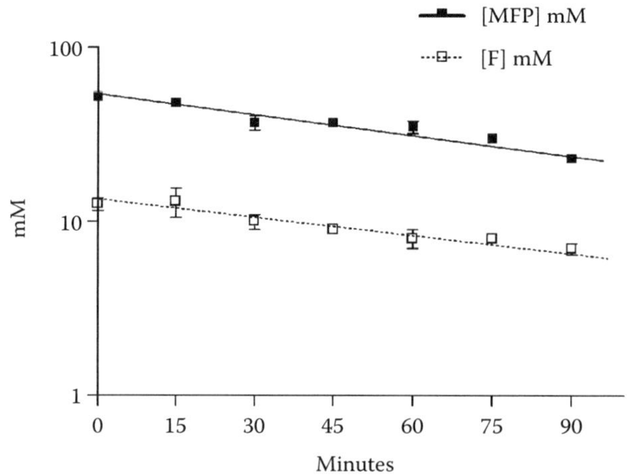

FIGURE 17.2 Changes of fluoride and MFP concentrations in the gastric lumen are shown.

REFERENCES

1. Manis, J., and D. Schachter. 1961. Active transport of iron by intestine: Features of the two steps mechanism. *Am. J. Physiol.* 203: 73.
2. Rigalli, A., M. Cabrerizo, A. Beinlich, and R.C. Puche. 1994. Gastric and intestinal absorption of monofluorphosphate and fluoride in the rat. *Arzneimittelforschung*. 44(I): 651–655.
3. Rigalli, A., A. Beinlich, and R.C. Puche. 2001. Intestinal absorption of fluoride at high luminal concentration of fluoride. *Arzneimittelforschung*. 51(I) :151–155.
4. Beinlich, A., L. Brun, A. Rigalli, and R.C. Puche. 2003. Intestinal absorption of disodium monofluorophosphate in rat as affected by concurrent administration of calcium. *Arzneimittelforschung*. 53(8): 584–589.
5. Rigalli, A., C. Aguirre, M. Armendáriz, and G. Cassiraga. 2003. Estudios de absorción en órganos aislados. In*: Formulación de modelos matemáticos de fenómenos biológicos*, 1st ed., Rosario, Argentina: Ed de la Universidad Nacional de Rosario, pp. 67–92.
6. Calderari, S., M.T. Font, O. Garrocq, S.M. Martínez, R.C. Puche, and M.C. Tarrés. 1991. The inbred IIM/FcM stock. *Rat Newsl.* 25: 28–29.

18 *In Situ* Isolation of the Intestinal Loop

María L. Brance, Lucas R.M. Brun, and Alfredo Rigalli

IN SITU ISOLATION OF THE INTESTINAL LOOP

Utility

This technique is applied to the study of the intestinal absorption of substances, their interaction with other constituents of the intestinal content, and the mechanism involved in their degradation in the intestinal lumen. It is a widely used technique and there are numerous writings that reveal the results obtained with it.[1-3]

This technique has an advantage in that the portion of intestine to be studied is not separated from the animal, which remains alive during the experiment. A disadvantage is that the organ cannot be studied completely and the concentration in the blood flowing through the intestine cannot be measured. This last inconvenience is overcome by the intestine perfusion, which is discussed in Chapter 20.

Materials

General anesthetics
Thermostatized rat board
Ring-handled preparation scissors
Ring-handled iris scissors
Cotton pliers (2)
Adson forceps (2)
Artery forceps (4)
Needle holder
Saline solution
Iodopovidone
Solution for filling intestinal lumen
1-ml syringe
Polyethylene catheter (PC-40 or similar)
Suture needle ½ – 5 to 10 mm
Suture thread
Cotton buds
Gauze

Procedure

Once the animal is under general anesthesia, it is shaved and the antisepsis of the incision area of the abdomen is performed while waiting for the analgesic and muscle relaxant effects. After completing a 2 to 3 cm median laparotomy, identification of the area of the intestine is done. It is necessary to locate the pylorus, which is in the upper area of the abdomen, under the left lobe of the liver, and after the stomach. This sphincter is clearly distinguishable because it has a ring that has a different color than the stomach and intestine, which are easier to recognize. The stomach is large, has a greater diameter, and has a grayish color with stretch marks, while the intestine is tubular and has a pink or reddish color. If the duodenum is to be studied, a double thread is passed under the pylorus with cotton pliers and a knot is made. The position of the second knot is measured from the position of the first knot, and the chosen distance depends on the experiment. On the second knot, a 45° cut is made in the middle of the intestine and a catheter is introduced through it. If possible, the catheter should have rounded ends and multiple lateral orifices. Also, the catheter must have a wide ring on its end to prevent it from sliding out during the experiment. The knot must be made on the previously mentioned ring (Figure 18.1). This ring can be easily made by winding firmly tied surgery thread onto the polyethylene catheter or as explained in Chapter 12.

The solution to be studied can be introduced through the catheter that has a syringe on the other end. Samples of the content of the lumen can be extracted through the same catheter. An artery forceps must be placed on the catheter to prevent the loss of the content of the intestinal lumen, so that when the syringe containing the sample is removed, the rest of the fluids of the lumen are not eliminated through the catheter.

Precaution: The threads, although firmly tied, must not put excessive pressure on the tissue, as this causes bleeding or, in extreme cases, blood clots in the intestinal lumen, which can hamper obtaining of reliable data. It is also convenient that, when injecting the solution, the walls of the intestine are not extremely tense, as this can affect circulation.

The amount of samples will depend on their volume as well as the size of the portion of isolated intestine. Approximately 0.5 ml of solution can be used per each centimeter of isolated intestine in a rat of 200 to 250 g of body weight.

In the case of nonfasted animals, it is possible to fill the lumen with solution, and extract and eliminate it. In this way, the rest of any food is eliminated, as determinations could be difficult as well as the extraction of samples. After that, the definitive solution is instilled in the lumen.

When the concentration of a substance in the intestinal lumen is determined, it is necessary to consider the possible change of volume due to gastrointestinal secretions or to the absorption of water produced in that area. If there is a substance that is not absorbed in the duodenum or has bound to its walls, its determination can be used as a control to the changes in volumes. An effective technique, when these volumes are not available, is to perform a summation of volumes. The volume and the concentration of the substance of the solution injected, as well as the initial concentration of the substance in the lumen, must be known. Then, notes are taken to specify the extracted volumes. When the experiment is finished, all the remaining liquid is extracted (residual volume) and its volume is measured using the weight and density of the intestinal content

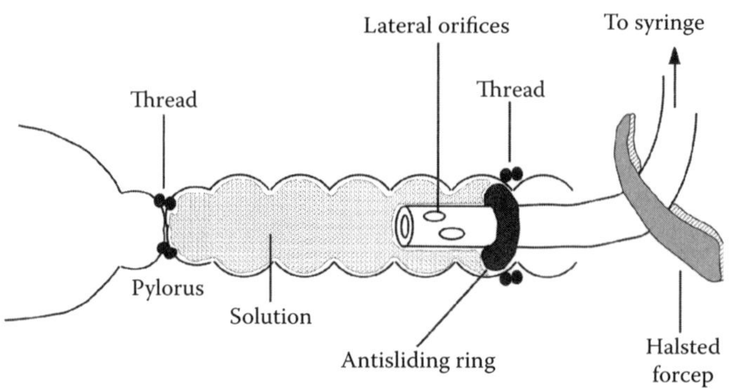

FIGURE 18.1 Placing of the distal catheter in the experiment of *in situ* isolation of intestinal loop is pictured.

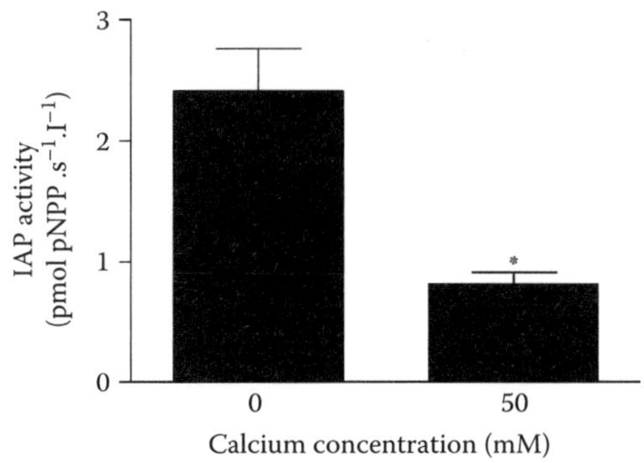

FIGURE 18.2 Shown is intestinal alkaline phosphatase activity in the intestinal lumen with and without calcium 50 mM after 20 minutes (mean ± SD). The asterisk (*) represents significant differences with respect to calcium 0 mM.

(usually 1 g/ml). If during the experiment there was not a great change in the volume through water absorption or secretion, the initial volume should be approximately the same to the sum of the extracted volumes and the residual volume at the end of the experiment.

Experimental Results

The objective of this work was to investigate in rats the effect of calcium on the activity of the intestinal alkaline phosphatase *in vivo*. Female adults rats (IIM/Fm) subline "m"[4] of 190 to 230 g of body weight were used. The rats were anesthetized with urethane (120 mg/100 g body weight) administered intraperitoneally. Room temperature was kept between 21 to 22°C during the experiment.

Four cm of the duodenum was isolated *in situ* by two knots, one of them at the pylorus level. The duodenum lumen of each rat of the control group (n = 4) was filled with 2 ml of saline solution (NaCl 9 g/l). The duodenum lumen of each rat of the "treated" group (n = 4) was filled with 2 ml of $CaCl_2$ 50 mM. The content of the intestinal lumen was extracted after 20 minutes following the placing of the solution. In the samples, the activity of the intestinal alkaline phosphatase (pmol pNPP.s^{1}.l^{-1}) was spectophotometrically measured, using p-nitrophenylphosphate (p-NPP) as substrate dissolved in diethanolamine 1 M pH 9.8. The activity of the alkaline phosphatase in the intestinal lumen was lower in rats treated with calcium 50 mM (Figure 18.2).

These results confirm the inhibitory effect of calcium on intestinal alkaline phosphatase.

REFERENCES

1. Manis, J., and D. Schachter. 1961. Active transport of iron by intestine: Features of the two steps mechanism. *Am. J. Physiol.* 203: 73.
2. Rigalli, A., A. Beinlich, and R.D. Puche. 2001. Intestinal absorption of fluoride at high luminal concentration of fluoride. *Arzneimittelforschung*. 51(I): 151–155.
3. Beinlich, A., L. Brun, A. Rigalli, and R.C. Puche. 2003. Intestinal absorption of disodium monofluorophosphate in rat as affected by concurrent administration of calcium. *Arzneimittelforschung*. 53(8): 584–589.
4. Calderari, S., M.T. Font, O. Garrocq, S.M. Martínez, R.C. Puche, and M.C. Tarrés. 1991. The inbred IIM/FcM stock. *Rat Newsl.* 25: 28–29.

19 Intestinal Everted Sacs

Lucas R. M. Brun, María L. Brance,
Lucas F. de Candia, and Alfredo Rigalli

INTESTINAL EVERTED SACS

Utility

This technique is used to study the intestinal absorption of substances, their interaction with other components of the intestinal content, the mechanism involved in its degradation in the intestinal lumen, or the pharmacological actions on the cells of the intestinal mucous. The experiment shares several characteristics with the *in situ* isolation of the intestinal loop described below.

The difference (as well as the disadvantage) from the *in situ* isolated intestinal loop is that the intestinal everted sacs are separated from the animal without keeping the irrigation and the innervation of the organ. However, it does allow for easier measurements of the luminal compartment because its volume can be modified according to experimental conditions.

Materials

General anesthetics
Ring-handled preparation scissors (2)
Ring-handled iris scissors
Adson forceps
Cotton pliers (2)
Thermostatic bath at 37°C
Measuring cylinder (adequate volume for the experiment)
Magnetic stirrer
3-ml syringe
Saturated potassium chloride solution for euthanasia
Saline solution
Iodopovidone
Filling solution for the everted sacs
Suture needle curved ½ – 5 to 10 mm
Suture thread
Cotton buds
Gauze

Procedure

Once the animal is under the effects of the anesthetic, euthanasia is performed using a saturated solution of potassium chloride through intracardiac injection (see Chapter 5).

A median laparotomy is performed and the following steps will depend on the portion of the intestine to be studied. The duodenum study will be used as the example here, but in case it is performed in the jejunum, ileum, or colon, the procedure is similar. The pylorus is located below the stomach, under the left lobe of the liver, and partially covered by the spleen. It is cut at that level and

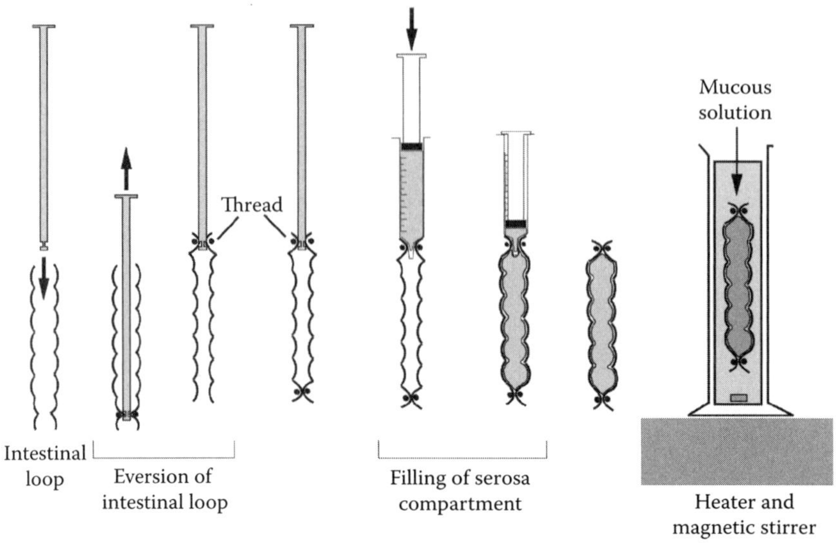

FIGURE 19.1 Preparation of duodenal everted sacs is shown. Thick arrows indicate movements, thin arrows indicate parts of the system.

carefully separated from its mesentery, hopefully avoiding an injury. Once this separation is completed, find a high resistance point (approximately 9 cm) where the end of the duodenum is marked due to the presence of the Treitz's ligament.

Then, a nonripper glass or plastic element approximately 3 mm in diameter is introduced into the duodenum segment. A plunger of a 1-ml syringe without the rubber end is adequate for this purpose. There should be a slot in the element, perpendicular to the major axis (Figure 19.1). On this area, a knot is made to adjust and introduce the intestine into that slot. Pulling softly from the end opposite to the knot can evert the intestine. In this way, the serous side ends up inside and the mucosa is outside of the everted sac.

Subsequently, the everted sacs are filled with a solution of known composition and, with the aid of an assistant, the sacs are closed. The length of the sacs depends on the experimental design.

The sacs are placed in a solution containing the metabolite to be analyzed. The sacs must be placed in a container at 37°C with constant stirring. Through measuring the concentration of the metabolite in the serous and mucus compartment, some conclusions about the mechanism of transport can be inferred.

Experimental Results

The objective of this work was to investigate the effect of luminal calcium (Ca) on the serum and luminal concentrations of intestinal alkaline phosphatase (IAP). Everted intestine experiments were carried out with three segments of 3 cm each from the pylorus, in the caudal direction.[1] Fasted female 200-g rats (strain IIM/Fm substrain "m") were used. The intestinal mucosa were exposed to the following solution (Smucosa): Tris, 1 mM; $MgCl_2$, 1 mM; glucose, 160 mM; Ca 1 mM, (Ca1); 10 mM (Ca10); or 100 mM (Ca100), and the serosal sides were exposed to the same solution without Ca (Sserosa). Samples were obtained from Smucosa at 0, 10, 20, 30, 50, 65, and 80 minutes, and from the Sserosa at the end of the experiment. IAP concentration was measured by Western Blot and expressed as integrated optical density units (IOD). Guinea pig anti-rat IAP primary antibody, antiguinea pig immunoglobulin G horseradish peroxidase-conjugated secondary antibody, and 3-amino-9-ethyl carbazole as substrate were used. Results are expressed as mean ± SEM. IAP concentration in Smucosa increased significantly along the experiment (Ca1: 4.3 ± 0.6, Ca10: 3.8

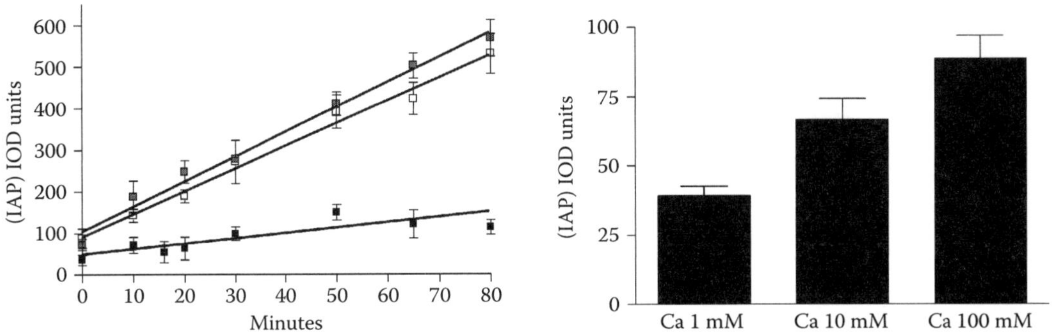

FIGURE 19.2 Left panel: Concentration of IAP in Smucosa; gray square: Ca1; white square: Ca10; black square: Ca100. Right panel: Concentration of IAP in Sserosa.

± 0.7, Ca100: 1.0 ± 0.2 IOD unit/min $p < 0.0001$, linear regression (Figure 19.2)). This increase followed an inverse relationship with Ca concentration ($p < 0.0001$). A direct relationship between IAP concentration in Sserosa and luminal Ca concentrations was observed (Ca1: 39.2 ± 3.5, Ca10: 66.7 ± 7.5, Ca100: 88.7 ± 8.2 IOD unit/min $p < 0.0001$ (Figure 19.2)). The conclusion was that the release of IAP to serosa and mucosa followed a direct and inverse relationship with Ca concentrations, respectively.

REFERENCE

1. Brun, L.R.M., M.L. Brance, L. de Candia, A. Traverso, and A. Rigalli. 2007. Rat intestinal alkaline phosphatase (IAP) levels at different intestinal calcium concentrations. Experiments with intestinal everted sacs. *Bone* 40(3): S8.

20 *In Situ* Perfusion of the Small Intestine

Alfredo Rigalli

IN SITU PERFUSION OF THE SMALL INTESTINE

UTILITY

This surgical manipulation is similar to the *in situ* isolation of the intestinal loop explained in Chapter 18. The technique allows one to simultaneously obtain data from the blood compartment and from the lumen of the isolated intestine. Comparable to the previously mentioned process, in this model, it is not isolated from the innervation of the area as well as the action of local factors from adjacent tissues. However, the fact that circulation is isolated in a perfusion process makes the intestine area inaccessible to the humoral systemic factors. As a consequence, possible metabolic regulations performed by these factors cannot be studied. One of the advantages of this model is that the concentration of substances in the intestinal lumen and in the buffer of perfusion can be measured simultaneously. This allows one to make a comparison of concentrations of substances in both compartments, inferring about their transportation mechanisms, which show associations to membranes or degradation while passing through the mucosa.

Materials

General anesthetic
Thermostatized rat board
Ring-handled preparation scissors
Ring-handled iris scissors
Cotton pliers (2)
Adson forceps (2)
Artery forceps (4)
Needle holder (1)
Infusion pump (piston or peristaltic type)
Three-way valve
Catheter PC-50 20 cm in length
Catheter PC-50 5 to 10 cm in length
Saline solution
Iodopovidone
Buffer of perfusion (according to each experiment)
Heparin
Filling solution for intestinal lumen
Suture needle curved ½ – 5 to 10 mm
Suture thread
Cotton buds
Gauze

Procedure

This surgical practice has two stages:

1. Catheterization of the upper mesenteric artery and portal vein, and the verification that the perfusion process is correct.
2. *In situ* isolation of the intestinal loop, as described in Chapter 18.

It is recommended that the practice should be performed in that order, so that, when isolating the intestine area, it is possible to know which sector is subjected to perfusion. When the perfusion process starts, the color of the tissue is lightened. If this does not happen, it is possible that a collapse in the vascular system has occurred preventing the buffer from arriving at the desired area.

Once the animal is anesthetized, a median laparotomy of 6 to 8 cm (depending on the size of the rat) is performed. Using two cotton buds, the small and large intestine are moved to the left of the animal until the vein cava, duodenum, and portal vein are visible (Figure 20.1).

The celiac artery is observed on the upper area, to the left of the vein cava, next to the liver. It is easy to distinguish as it presents a curvature and the characteristic pulsation of the arteries. Its size ranges between 3 and 5 mm. The upper mesenteric artery is located 1 cm below the celiac artery, perpendicular to the vein cava and leading to the left. It is straight and has a pulsation, although less evident than the one of the celiac artery.

Catheterization of the Upper Mesenteric Artery

The upper mesenteric artery must be isolated using the conventional arterial catheterization technique (see Chapter 13). Briefly, a double thread is placed under it and two knots are made on it (one proximal and the other one distal) that will be tightened once the catheterization is performed. The thread also must be passed under the celiac artery, making a knot that will be tightened at the end of the process.

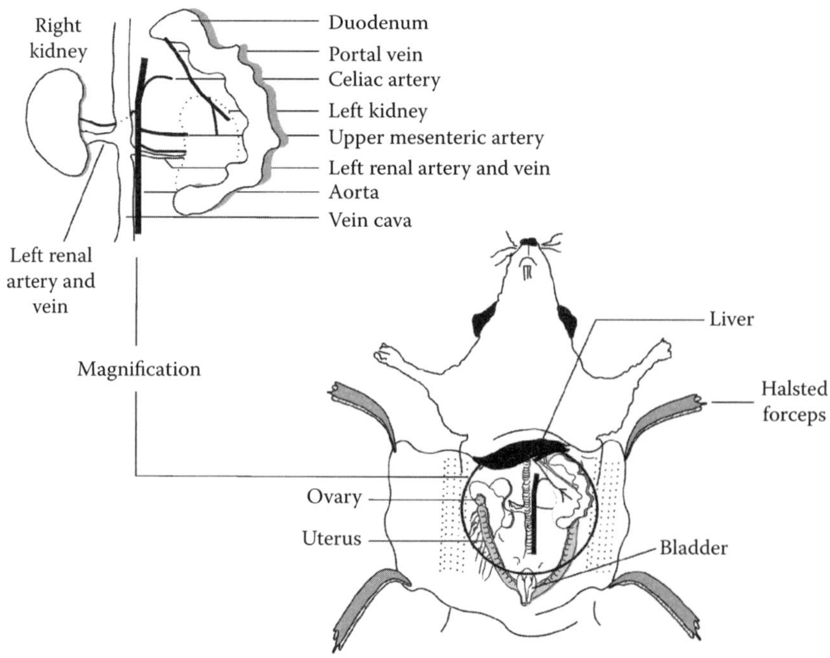

FIGURE 20.1 Operative field for the intestine perfusion.

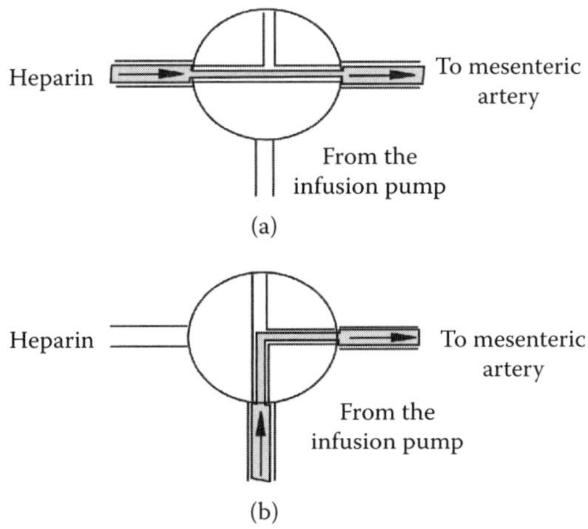

FIGURE 20.2 Handling of three-way key for intestine catheterization.

The proximal knot on the mesenteric artery is tightened and, placing a spatula under the artery, a small orifice is made with a 15 × 0.5 injection needle. The bevel of the catheter is introduced in this orifice. The catheter must be drained with buffer and connected to a syringe with 0.5 ml of heparin (Chapter 45), through a three-way valve (Figure 20.2a/b). At the moment of placing the catheter, the contents of the syringe are injected in order to heparinize the area that will be perfused (Figure 20.2a), the valve is changed, and it starts to perfuse with a flow of 0.1 to 0.2 ml/min (Figure 20.2b). After verifying that there is no buffer leaking from the catheter, the distal knot of the mesenteric artery is tightened to the catheter.

Catheterization of the Portal Vein

Although the portal vein's dimensions can create the illusion that this is an easy task, the catheterization of this vein is difficult due to its thin walls and, it should be placed on a serous, not on a firm base, as happens with the femoral and mesenteric artery. A double thread and a spatula must be placed under the vein and an orifice made with a 15 × 0.5 injection needle. The orifice is performed in the direction opposite to the blood flow, and the catheter placed in the same direction. Once the vein is catheterized, it is bound with the thread in order to fix the catheter to the vein.

After verifying that the blood is moving up through the catheter, the celiac artery is tied to prevent the blood from entering the perfused area. Once this procedure is performed, if the surgery was successful, the liquid obtained must be limpid or with only a small amount of cells that can be eliminated through centrifugation.

In Situ Isolation of the Intestinal Loop

Once it is verified that the perfusion is correct and that the best perfused area of the intestine is identified, the isolation of the area of the intestine is performed, as described above.

Once the experiment is finished, a 5% w/v methylene blue solution can be perfused. This colorant dyes the perfused area and verifies the success of the surgery, guaranteeing the validity of the results.

Special Considerations

This solution can be a buffer KRB (Chapter 45) to which albumin is added to maintain the osmotic pressure of the fluid and prevent the edematization of the tissue. Another possibility is the perfusion with plasma extracted from animals of the same strain or whole blood. The latter has the advantage of a better oxygenation of the tissue.

The perfusion can be done using a sachet containing the solution and a catheter of the adequate length that allows applying the correct pressure to the system to obtain a convenient flow.

Section V

Pancreas

21 Experimental Models for the Study of Diabetes

Verónica E. Di Loreto and Alfredo Rigalli

INTRODUCTION

The rat pancreas is a dispersed soft tissue organ and can be mistaken for adipose tissue. It can be found in the proximity of the duodenum and close to the spleen. Despite its similarity to the adipose tissue, the pancreas has two important characteristics that allow its identification. Whereas the adipose tissue is bright white, the pancreas has a grayish color and is less shiny. The most significant difference is the density of the tissue. The adipose tissue is less dense than water, therefore, it floats; a phenomenon not observed with the pancreas. It is suggested that the surgeon extract both tissues, put them in saline solution, and check the density difference mentioned above.

The pancreas is an organ with exocrine and endocrine secretion. To achieve this function, the tissue has two different kinds of glands: the pancreatic acini, responsible for exocrine secretion; and the islets of Langerhans, responsible for the endocrine secretion. The exocrine secretion is composed of a fluid that flows into the duodenum and contains, apart from electrolytes, enzymes responsible for the nutrients digestion. This secretion is stimulated by gastrointestinal enzymes, such as secretin and pancreozymin. The first is released by cells of the duodenum mucus stimulated by the chime with acid properties. It also stimulates a rich bicarbonate secretion with poor enzymatic activity. The pancreozymin, also known as cholecystokinine, is released by the duodenum cells stimulated by lipids and carbohydrates. This hormone stimulates a secretion that is rich in pancreatic enzymes and their zymogens. The endocrine secretion is composed of insulin, glucagon, and somatostatin, produced by the beta, alpha, and delta cells from the islets of Langerhans, respectively.

Therefore, it is inferred that the pancreas should be considered in all the experiments relating to nutrition, digestion, glucose homeostasis, and triacylglycerol metabolism.

A large number of animals from different ranges and species have been used as models to study diabetes.

The term *diabetes animal model* is frequently used when experiments with diabetic animals are performed, with their meaning extrapolated to human diabetes. A "model" does not mean a reproduction of human diabetes on an animal, and it does not represent the complete range of aberrations or complications observed in human diabetes either. The potential of an animal model lies in the chance to explore specific variables (morphologic, biochemical, immunologic, metabolic) or therapeutic measures (drugs, transplants) that are unable to be performed on humans. The conclusions should be confined to the pathogenesis of a particular disorder or its correction because there is no total equivalence to the human disease.[1]

SURGICAL PANCREATECTOMY

Although this is not the most convenient method for diabetes induction, it has been used through many years with good results for that purpose. To induce diabetes, 95% of the pancreas must be removed. Diabetes develops after 3 months.[2]

Materials

General anesthetic
Thermostatized rat board
Ring-handled preparation scissors
Ring-handled iris scissors
Cotton pliers (2)
Needle holder
Iodopovidone
Suture needle curved ½ – 5 mm
Suture thread
Cotton buds

Procedure

Once the animal is under anesthetics, a paramedian left laparotomy is performed. The spleen is identified and the tissue next to it is held and lifted. In this way, the pancreas is spread out and the vessels irrigating the pancreas are observed. These vessels are bound so that when the ablation of the pancreatic tissue is performed, important hemorrhages do not occur. The portion in contact with the duodenum is hard to extirpate and this must be done very carefully because the pancreobiliar duct goes through it, and, if it is cut, it will cause hepatic dysfunction, which is related to the elimination of substances through this duct, and with lipids digestion.

The inconveniences of this kind of surgery lie in the double function of pancreas that has influence one on the other. If one wishes to perform experiments related to the insulin and/or glucagon action or to the metabolisms that these hormones control, remember that the process of degradation and absorption of nutrients will be affected. The considered inconveniences of this surgery can be solved by the use of chemical induction of diabetes.

CHEMICAL INDUCTION OF DIABETES

Several chemical agents are selective toxins for the beta cells of the pancreas. The main chemical agents include: alloxan, a pirimidine that presents a structural similarity to uric acid (2,4,5,6 tetraoxohexahydropirimidine), and streptozotocin (STZ) (2-deoxy-2- (3-methyl-3-nitrosoureido)-D-glucopyranose), which can be considered a glucose with a lateral nitrosourea chain with a high reactivity.[1] Both produce a complete or almost complete destruction of the beta cells by direct toxic effects.[2,3]

As was emphasized previously, even though an animal model cannot be extrapolated to human diabetes, it is commonly accepted that this chemical induction model is similar to human type I diabetes.

Alloxan

Alloxan (2,4,5,6 tetraoxohexahydropirimidine) has a complex electronic structure and there are several tautomeric forms. Its structure is shown in Figure 21.1.

It is highly unstable in neutral pH water, and reasonably stable in a solution with pH lower than 3.[1] There is evidence that alloxan's diabetogenic action is voided approximately 5 or 10 minutes after the intravenous or intraperitoneal injection because alloxan is unstable in pH physiologic.[3] Alloxan administration can be performed through intravenous injection in a 40 to 45 mg/kg of body weight dose[1] or using intraperitoneal doses between 50 and 200 mg/kg of body weight.[4] The plasma glucose value on a nonfasted normal rat is 90 to 160 mg/dl, with water and food *ad libitum*. With a dose of intraperitoneal alloxan of 200 mg/kg, dissolved in a citrate–phosphate buffer (pH 4)

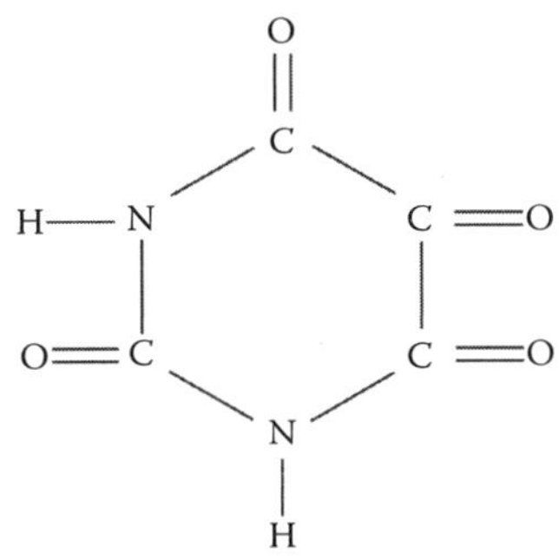

FIGURE 21.1 The chemical structure of alloxan.

with a 20-g/100 ml concentration, the glucose levels would be 150 to 900 mg/dl in 48 hours.[5] The seriousness of the diabetes induced by intravenously injected alloxan is the same as if it were intraperitoneally injected.[4] This produces an irreversible functional damage to the pancreatic beta cells in a few minutes and structural changes in hours on most rodents, dogs, cats, rabbits, apes, sheep, fish, and birds.[1] To develop diabetes, it is necessary to destroy or cause an inactivation of more than 95% of the islets.[4]

The response to alloxan can be divided into three phases: initial hyperglycemia due to hepatic glycogen degradation, which lasts approximately 2 hours; temporary hypoglycemia in the next 6 hours, caused by the secretion of insulin from the damaged cells; and, finally, permanent hyperglycemia that begins in 12 hours.[1]

The percentage of diabetic rats obtained a week after the alloxan administration depends on the rat sensitivity to the drug. Alloxan is a kidney tubular cell toxin, in particular, when high doses are administered and this is the most likely cause of animal death.[6] The animal that survives 5 days following the alloxan injection usually presents diabetes with variable seriousness characterized by polyphagia, with a marked loss of weight,[4,5] polyuria and glycosuria, but without significant ketonemia or ketonuria.[5] Although in some projects ketonemia and ketonuria are described during the first week, these signs then disappear. The explanation for this phenomenon is that the sudden insulin deficiency allows a fast catabolism and mobilization of triacylglycerides. When endogenous lipids run out, ketosis disappears.[7]

On experiments made in our laboratory with strain IIM/Fm subline "m" male rats, we observed that 50% of the animals die the week following alloxan administration. Surviving animals presented a wide range of blood glucose levels (200 to 900 mg/dl) and a normal acid-base state.[8]

Metabolic changes on alloxan-induced diabetes are similar to other insulin deficiency conditions. Histologically, on beta cell levels, a degradation of the cell's membrane is observed, on the endoplasmic reticulum and mitochondria in 2 hours, followed by nuclear and cellular degradation.[6]

Pathogenesis

Although the diabetogenic activity of the alloxan has been known for approximately 60 years, the pathogenic mechanism has only recently become clear. Alloxan produces reactive oxygen species and the DNA of pancreatic islets is one of the targets.[9] As a consequence, nicotinamide adenine dinucleotide (NAD^+) decreases due to damage of DNA. In addition, alloxan has a high affinity to SH-containing cellular compounds and, as a result, reduces glutathione content. Furthermore, alloxan produces the inhibition of the glucokinase, an SH-containing protein essential for insulin secretion induced by glucose.[6] Altered intracellular calcium metabolism and poor production of adenosine triphosphate (ATP) are also effects of alloxan.[3]

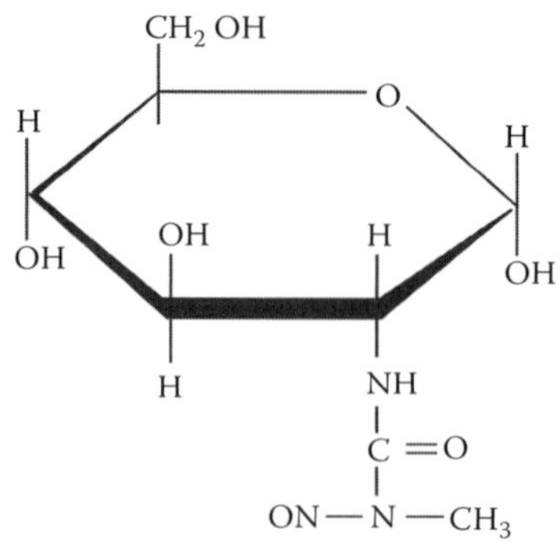

FIGURE 21.2 The chemical structure of streptozotocin.

Streptozotocin

Streptozotocin (2-deoxy-2-(3-methyl-3-nitrosoureido)-D-glucopyranose) (Figure 21.2) is an unstable drug even in an acid pH solution and must be injected immediately after being dissolved in a pH 5 citrate buffer. The half-life *in vivo* is shorter than 15 minutes. Streptozotocin (STZ) is administered usually in a unique 40 to 90 mg/kg body weight dose intravenously or intraperitoneally.[1] It is effective on different species, such as rats, dogs, mice, apes, and rabbits. There is a high risk of mortality during the first 48 hours after the administration of the unique diabetogenic dose. To avoid this mortality risk and the damage associated with a unique dose of this drug, many low doses of STZ can be used to reach a moderated chronic hyperglycemia level (300 to 600 mg/dl) without requiring insulin.[10]

This repeated STZ administration, with small doses insufficient for an immediate diabetogenesis, produces insulitis, which leads to an almost complete destruction of the beta cells, similar to that observed in type I diabetes patients. As a result of the *in vivo* action of STZ, 4 days after the administration, the beta cells appear as degranulated, although not necrotic, and there are signs of limitation in their proliferation. No transformation of ductal or acinar cells on beta cells is observed, showing that any beta cell proliferation, after the first attack, is due to preexistent precursors of the beta cell's progeny. Although it is not nephrotoxic like alloxan, some long-term studies show that it can induce kidney tumor growth.[2]

Pathogenesis

The toxicity over beta cells would be related to its nitrosourea configuration while the deoxyglucose part would make its transportation through the cellular membrane easier.[1]

The mechanism through which the STZ produces its cytotoxic effect is not completely clear. The STZ reduces the NAD^+ content in several tissues and its effect is particularly harmful and necrotizing on pancreatic beta cells.[1] It acts mainly by producing alkylation of DNA.[6] The beta cells destruction is probably the consequence of low NAD^+ levels caused by the nuclear poly (ADP-ribose) synthetase during DNA reparation.[3,11] The low levels of NAD^+ also produce a decrease in intracellular ATP levels.[3,12]

On the other side, the participation of nitric oxide and reactive oxygen species in the citotoxic effects of STZ was proposed.[6]

As a conclusion, Okamoto[12] proposes a common mechanism of action for alloxan and STZ toxicity suggesting that the beta cell, trying to repair the damaged DNA, produces a suicidal response. The initial injury is produced by different causes: alloxan acts mainly by production of reactive oxygen species and STZ by DNA alkylation.

REFERENCES

1. Shafrir, E. 1990. Diabetes in animals. In *Diabetes mellitus, theory and practice*, 4th ed. Eds. Harold Rifkin and Daniel Porte Jr. Amsterdam: Elsevier, pp. 299–340.
2. Kempe, H.P., K. Engelmann, N. Greta, and C. Hasslacher. 1993. Models of diabetes for studying diabetic nephropathy. In *Experimental and genetic rat models of chronic renal failure*. Basel: Karger, pp. 148–155.
3. Boquist, L. 1992. Aspects of the diabetogenicity of alloxan and streptozotosin, with special regard to a mitochondrial hypothesis. In *Lessons from animal diabetes*, vol iv. Ed. E. Shafrir. Biennial Review Volume, Great Britian: Smith-Gordon, pp. 1–16.
4. Lazarow, A., and S. Palay. 1946. The production and course of alloxan diabetes in the rat. *J. Lab. Clin. Med.* 31(9): 1004–1015.
5. Locatto, M.E., H. Abranzon, D. Caferra, M.C. Fernández, R. Alloatti, and R.C. Puche. 1993. Growth and development of bone mass in untreated alloxan diabetic rats. Effects of collagen glycosilation and parathyroid activity on bone turnover. *Bone Miner.* 23: 129–144.
6. Szkudelski, T. 2001. The mechanism of alloxan and streptozotocin action in B cells of the rat pancreas. *Physiol. Res.* 50: 536–546.
7. Steiner, D., V. Rauda, and R. Williams. 1961. Severe ketoacidosis in the alloxan diabetic rat. *Endocrinology* 68(5): 809–817.
8. Locatto, M.E., V. Di Loreto, M.C. Fernández, D. Caferra, and R.C. Puche. 1997. Relative weight of glucosa, insulin and parathyroid hormones in the urinary loss of phosphate by chronically diabetic rats. *Acta Diabetol.* 34: 211–216.
9. Malaisse, W. 1982. Alloxan toxicity to the pancreatic B-cell. *Biochem Pharmacol.* 31(22): 3527–3534.
10. McNeill, J.H., Ed. 1999. *Experimental models of diabetes.* Boca Raton, FL: CRC Press, pp. 257–295.
11. Sandler, S., and I. Swenne. 1983. Streptozotocin, but not alloxan, induces DNA repair synthesis in mouse pancreatic islets in vitro. *Diabetologia.* 25: 444–447.
12. Okamoto, H. 1996. Okamoto model for beta cell damage: recent advances. In *Lessons from animal diabetes*, vol. vi. Ed. E. Shafrir. Boston: Birkhaussen, pp. 97–111.

22 Islets of Langerhans Isolation

Alfredo Rigalli

ISLETS OF LANGERHANS ISOLATION

Utility

This technique is used in *in vitro* experiments and allows for the isolation of the islets from the exocrine secretion of the pancreas.

Materials

General anesthetic
Thermostatized rat board
Ring-handled preparation scissors
Ring-handled iris scissors
Cotton pliers (2)
Adson forceps
Rat-tooth forceps
Pasteur pipette
Latex tube with nozzle
Petri dish (2)
Collagenase
Stereoscopic magnifying glass 5×, 10×, 20×
Ice
Saline solution
Iodopovidone

Procedure

Once the rat is euthanized, a pancreatectomy is performed. It is recommended to use the part of the pancreas that is close to the spleen because it is easier to access. The tissue obtained is washed twice with a Krebs–Ringer bicarbonate buffer (KRBB). Glucose must not be added to this buffer to avoid the insulin secretion stimulus.

The tissue is cut in small pieces of 1 to 2 mm and digestion with collagenase is performed.[1] If the rat weighs approximately 200 g, the pancreas is incubated with 16 mg of collagenase in the KRBB buffer, at 37°C. The content of the test tube is shaken lightly during the first minute, and at 1 to 2 minute intervals a drop is extracted and observed with a magnifying glass with a magnification of 10× to 20×. The islets look like white flakes intermingled with exocrine acinar tissue. Using a Pasteur pipette joined to a latex tube with a nozzle, the islets are isolated. When it is confirmed that the islets are free from the rest of the tissue, hydrolysis is inhibited by adding the same volume of KRBB buffer at 4°C. With this maneuver, the enzyme is diluted and the lower temperature decreases its action. The first isolation is performed. A drop of the incubated pancreas is placed on a Petri dish and, with the help of a magnifying glass and a Pasteur pipette, the islets are separated and placed on another Petri dish with KRBB. It is necessary to carefully draw up through the pipette in order to isolate the islets without obtaining exocrine tissue. Then, from the dish where the islets are placed, a new isolation is performed paying more attention to the elimination of exocrine acinar tissue.

The islets obtained can be used for incubation or perifusion techniques. During incubation, the islets are placed in a buffer solution, and samples are extracted from the system when necessary. Hormone accumulates in the buffer throughout the experiment. On the other hand, during perifusion, the islets are placed over a layer of Sephadex® (Pharmacia Fine Chemical, Uppsala, Sweden), which acts as a bed; buffer flows through Sephadex and is collected in fractions of adequate volume. Each fraction contains the hormone that has been produced during this period of time.

Both techniques can provide nearly the same information. However, perifusion allows exposure of the tissue to glucose in a stimulatory concentration and then exposure again to basal concentrations. Perifusion also allows exposure of the islet to different substances, such as glucose, over a period of time, modification of its concentration, or its removal from the system.

Experimental Results

The orogastric administration of sodium fluoride produces a reduction in plasma insulin concentration. Perifusions of isolated islets of Langerhans were performed with the aim of identifying whether this decrease in insulinemia is due to a reduction in the hormone secretion.[2] Fifty islets of Langerhans were subjected to perifusion with a KRBB buffer to a 0.4 ± 0.1 ml/min rate. The buffer had a glucose concentration of 0.6 g/l after 20 minutes and 3 g/l after 30 minutes. Sodium fluoride was added to the perifusion buffer to obtain a concentration of 0, 5, 10, and 20 μM. The samples were collected in periods of 20 (glucose 0.6 g/l) and 30 (glucose stimulatory period, 3 g/l) minutes. The insulin productions of 50 islets with different fluoride concentrations were compared. Insulin was determined using a radioimmunoassay (RIA) and the values obtained are shown in Table 22.1.

As can be seen, insulin production is higher with glucose 3 g/l and it is inhibited according to the fluoride concentration in the incubation medium. Similar results were obtained incubating 50 islets of Langerhans in a 1 ml KRBB buffer volume.

TABLE 22.1
Insulin Production (μIU/min 50 islets)

	Glucose 0.6 g/l	Glucose 3 g/l
Control	6.5 ± 0.1	15.9 ± 3.2
Fluoride 5 μM	1.3 ± 0.1	12.2 ± 0.8
Control	4.8 ± 0.5	26.3 ± 2.4
Fluoride 10 μM	0.8 ± 0.1	16.9 ± 1.0
Control	4.7 ± 0.1	15.6 ± 1.9
Fluoride 20 μM	0.5 ± 0.1	5.3 ± 0.6

Note: The results are expressed as the mean ± SD.

REFERENCES

1. Lacy, P.E., and F.M. Kostianovsky. 1967. Method for the isolation of intact islets of Langerhans of the rat pancreas. *Diabetes*. 16: 35–39.
2. Rigalli, A., J.C. Ballina, E. Roveri, and R.C. Puche. 1990. Inhibitory effect of fluoride on the secretion of insulin. *Calcif. Tiss. Int*. 46: 333–338.

23 Incubation of Pancreatic Tissue Slices

Inés Menoyo and Alfredo Rigalli

INCUBATION OF PANCREATIC TISSUE SLICES

Utility

This technique can be used as an alternative to the incubation of isolated islets of Langerhans. The most important difference is that in this technique, exocrine pancreatic tissue coexists with islets of Langerhans, and the production of proteases can decrease the concentration of insulin in the incubation buffer.

Materials

General anesthetic
Thermostatized rat board
Ring-handled preparation scissors
Ring-handled iris scissors
Cotton pliers (2)
Adson forceps
Rat-tooth forceps
Petri dish (2)
5-ml syringe
10-ml syringe (2)
Injection needle 15 × 0.5
Polyethylene catheter
Thermostatized chamber
Carbogen
Ice
Saline solution
Iodopovidone

Procedure

When pancreatic tissue is required, it is preferable to use the cervical fracture method instead of an excess of anesthetic to produce the death of the animal. After killing the rat, a median laparotomy is performed and the pancreas is infiltrated with 5 ml of Krebs–Ringer bicarbonate buffer (KRBB) at 4°C. When making this maneuver, the tissue will separate leaving areas with the solution. The objective of this treatment is to cool the tissue and produce a decrease of the enzymatic activity. To perform this procedure, a 15 × 0.5 injection needle must be used. After doing this, the pancreas area that is next to the spleen is extracted and placed in a 10 ml KRBB buffer at 4°C. The pancreatic and fat tissues are separated with curved, ring-handled preparation scissors and cotton pliers.

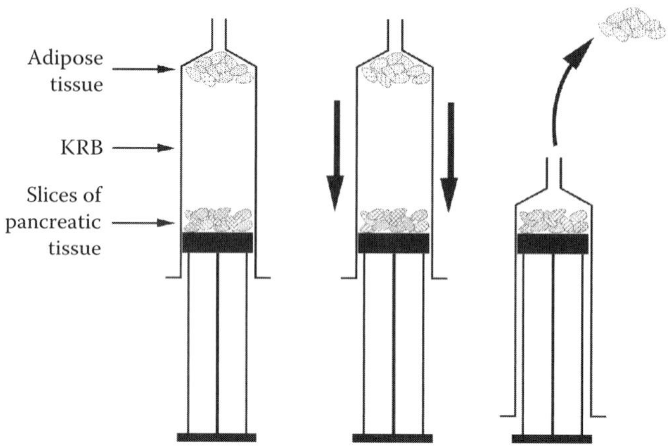

FIGURE 23.1 Pictured is the separation of pancreas slices from the adipose tissue.

Subsequently, the pancreatic tissue is transferred to a Petri dish containing approximately 2 ml of KRBB and cut in small slices (no larger than 1 mm) with ring-handled iris scissors. Once this has been achieved, the buffer, along with the slices, are drawn up with a 10 ml syringe (without needle) and left to rest for 5 minutes in a vertical position over the plunger (Figure 23.1).

Thanks to this procedure, it is possible to separate the pancreatic from the adipose tissue. The latter remains in the upper part of the syringe, while the pancreatic tissue is placed over the syringe plunger because of its higher density. Keeping the syringe in a vertical position, the liquid starts to be eliminated, taking with it the adipose tissue. The slices are suspended again in the same 1:1 buffer proportion and the procedure is repeated three to five times with KRBB buffer at 4°C. Finally, the remaining floating elements are eliminated and the slices are kept for subsequent work. To eliminate the buffer and the adipose tissue, a catheter with a diameter similar to that of the syringe end can be added to the syringe.

For incubation, the pancreatic tissue slices are placed at 37°C in the quantities and periods of time required for each experiment, gassed with carbogen (95% O_2, 5% CO_2) in an incubation buffer with the adequate composition for the experiment.

For an incubation chamber, a reversed syringe can be used, putting the carbogen into it through its lower side (Figure 23.2). The addition of carbogen not only maintains the pressure of oxygen, but also mixes the system. The syringe must be placed in a container at 37°C.

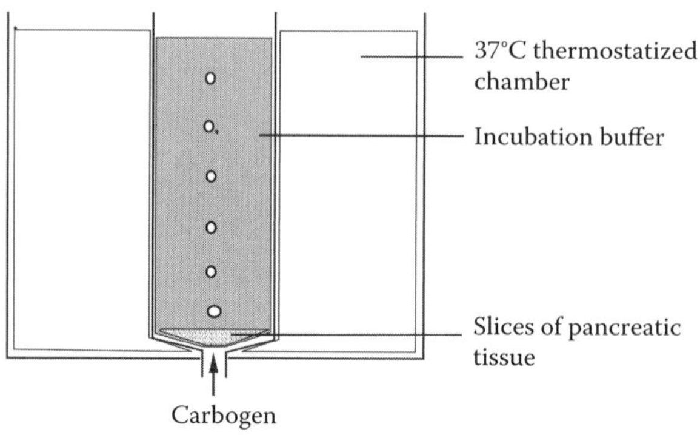

FIGURE 23.2 Pancreas slices incubation chamber is shown.

Experimental Results

To begin with, a short description of the insulin secretion mechanism is recorded. The glycemia increase leads to a biphasic pattern of insulin secretion. The initial phase of the response is mediated by phenomena from the calmodulin branch[1] and the maintained phase is supported by events from the protein kinase C from the calcium messenger system.[2] The two phases of the insulin secretion can be activated separately: (1) the branch where the calmodulin takes part through the divalent ionophore A23187,[3] and (2) the branch where the protein kinase C takes part through the phorbol ester 12-0-tetradecanoil-phorbol-13-acetate (TPA).[3,4] The ionophore A23187 induces the initial peak of the insulin secretion comparable to the one induced by glucose, while the phorbol ester induces the progressive increase of the second phase of the secretion similar to the one induced by glucose. The forskolin[5] that is an activator of the adenylate cyclase enzyme induces both phases of the insulin secretion. The plasma concentration of insulin decreases significantly after the administration of an oral dose of sodium fluoride (NaF) on rats and human beings.[6]

Incubations of 200 mg of pancreatic tissue slices were performed with the buffer described above. NaF was added to obtain a final concentration of 20 μM.[7] After 10 minutes of incubation, the insulin secretion was stimulated with:

12-0-tetradecanoil-phorbol-13-acetate (TPA) 100 nM
Ionophore A23187 2.5 μM
Forskolin 25 μM together with glucose 3 g/l[8]
Glucose 3 g/l

Simultaneously, incubations with the same stimuli of insulin secretion with no addition of NaF in the incubation medium were performed.

In both cases, after 50 minutes from the stimulus of the insulin secretion, the concentration of insulin and fluoride on the supernatant was determined. The experimental groups were:

Treatment B: incubation for 60 minutes in control conditions
Treatment F: incubation for 60 minutes in presence of 20 μM NaF
Treatment G: incubation for 10 minutes, stimulation with glucose, and incubation for 50 minutes after the stimulation
Treatment F + G: incubation for 10 minutes in presence of NaF 20 μM, stimulation with glucose, and incubation for 50 minutes after the stimulation
Treatment FK + G: incubation for 10 minutes, stimulation with forskolin and glucose simultaneously, and incubation for 50 minutes after the stimulation
Treatment F + FK + G: incubation for 10 minutes in presence of NaF 20 μM, stimulation with forskolin and glucose simultaneously, and incubation for 50 minutes after the stimulation
Treatment A23187: incubation for 10 minutes, stimulation with ionophore A23187, and incubation for 50 minutes after the stimulation
Treatment F + A23187: incubation for 10 minutes in presence of NaF 20 μM, stimulation with A23187, and incubation for 50 minutes after the stimulation
Treatment TPA: incubation for 10 minutes, stimulation with 12-0-tetradecanoil-phorbol-13-acetate, and incubation for 50 minutes after the stimulation
Treatment F + TPA: incubation for 10 minutes in presence of NaF 20 μM, stimulation with 12-0-tetradecanoil-phorbol-13-acetate, and incubation for 50 minutes after the stimulation

The experiments were performed on 7-week-old female rats, IIM/Fm strain,[9] subline "m" weighing 167 ± 19 g. The animals were fed with a balanced chow *ad libitum*. This food contains

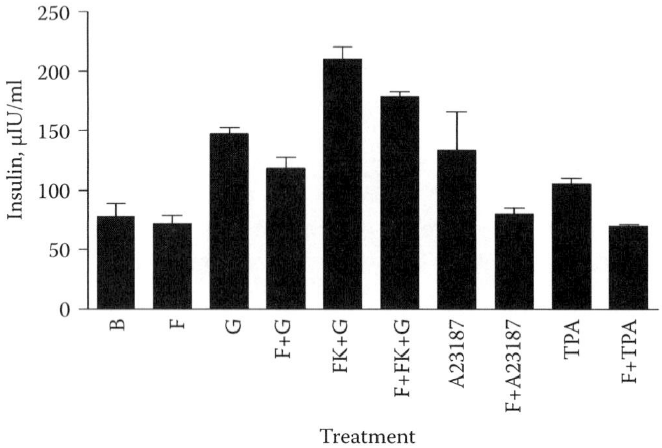

FIGURE 23.3 Insulin production by pancreas slices *in vitro* is shown.

2.4 μmole of fluorine/g. The animals drank tap water, which has fluoride in a concentration smaller than 15 μM. The pancreatic slices were incubated at 37°C, in the quantities and times indicated for each group. The incubation buffer with the following composition: NaCl 115 mM, KCl 4.7 mM, $CaCl_2$ 2.56 mM, $MgSO_4.7H_2O$ 1.2 mM, KH_2PO_4 1.2 mM, $NaHCO_3$ 25 mM, albumin 1% w/v, glucose 0.6 g/ l, pH: 7.4, was gassed with carbogen (95% oxygen, 5% carbon dioxide) throughout the experiment. Results are shown in Figure 23.3, and expressed as mean ± SD. The results are taken from the average of five experiments. The treatments were significantly different ($p < 0.0001$). The data show that insulin secretion is inhibited by fluoride, independently of the mechanism implicated in the secretion. The fluoride would affect a final stage of the secretory process, common to the mechanisms that imply calcium, AMPc, and protein kinase C.

REFERENCES

1. Rasmussen, H. 1986. The calcium messenger system (first of two parts). *N. Eng. J. Med.* 17:.1094–1101.
2. Rasmussen, H. 1986. The calcium messenger system (second of two parts). *N. Eng. J. Med.* 18:.1164–1170.
3. Zawalich, W., Ch. Brown, and H. Rasmussen. 1983. Insulin secretion: Combined effects of phorbol ester and A23187. *Biochem. Biophys. Res. Commun.* 117:.448–455.
4. Aizawa, T., M. Komatsu, N. Asanuma, Y. Sato, and G.W.G Sharp. 1998. Glucose acting "beyond ionic events" in the pancreatic ß cell. *Trends Pharmacol. Sci.* 19: 496–499.
5. Erhardt, P.W. 1990. In search of the digitalis replacement. *J. Med. Chem.* 30:.231–237.
6. Rigalli, A., J.C. Ballina, E. Roveri, and R.C. Puche. 1990. Inhibitory effect of fluoride on the secretion of insulin. *Calcif. Tiss. Int.* 46:.333–338.
7. Menoyo, I., A. Rigalli, and R.C. Puche. 2005. Effect of fluoride on the secretion of insulin in the rat. *Arzneimittelforschung*. 55(8): 455–460.
8. Seamon, K.B. . Forskolin and adenylate cyclase: New opportunities in drug design. *Ann. Rep. Med. Chem.* Chap. 29. National Center for Drugs and Biologics, FDA, Bethesda, MD.
9. Calderari, S., M.T. Font, O. Garrocq, S.M. Martínez, R.C. Puche, and M.C. Tarrés. 1991. The inbred IIM/Fm stock. *Rat Newsl.* 25: 28–29.

24 *In Situ* Perfusion of the Pancreas

Inés Menoyo and Alfredo Rigalli

IN SITU PERFUSION OF THE PANCREAS

Utility

This methodology is used to investigate the insulin production or some other hormone produced by the pancreas under the action of the agonists or antagonists of the secretion of these hormones. The advantage of this technique, as regards the incubation of slices or islets of Langerhans, is that, during the experiment, the integrity of the animal is maintained.

Materials

General anesthetic
Ring-handled preparation scissors
Ring-handled iris scissors
Cotton pliers (2)
Adson forceps
Artery forceps (2)
Rat-tooth forceps
Needle holder
Infusion pump: syringe or peristaltic
Three-way valve
Catheter PC-50, 20 cm in length
Catheter PC-50, 50 cm in length
Saline solution
Iodopovidone
Perfusion buffer
Heparin
Suture needle curved ½ – 5 mm
Suture thread
Cotton buds
Gauze

Procedure

The rat must be anesthetized with urethane or any anesthetic that allows a prolonged surgical period. First, a laparotomy is performed, then pancreas perfusion is performed through a catheter placed in the celiac artery, collecting samples though another catheter placed in the portal vein. The upper mesenteric artery is closed to avoid the contamination of the perfusion buffer with blood from general circulation. Using a three-way valve in the celiac artery catheter at the beginning of perfusion, heparin is injected through the same catheter. Perfusion is performed using a peristaltic or piston pump, maintaining a pressure of 80 to 110 mmHg. The samples are collected through the portal vein catheter in volumes that are adequate for the determinations to be performed. The perfusion flow must be maintained at between 0.1 to 0.3 ml/min.

Special Considerations

At the end of the experiment, perfusion with 1% w/v methylene blue must be performed in order to verify the perfusion extension.

If there are no pumps available, it can be done using a sterile solution bag, with the solution and a long catheter that allows to exert the right pressure to the system to reach the appropriate flow.

The buffer of perfusion can be KRB (Chapter 45) to which albumin is added to avoid edematization of the tissue. Another possibility is the perfusion that is autologous with plasma or blood. This one has the advantage of better oxygenation of the tissue.

Experimental Results

Four rats were perfused at an average flow of 0.13 to 0.21 ml/min with buffer pH: 7.4 that contained: NaCl 115 mM, KCl 4.7 mM, $CaCl_2$ 2.56 mM, $MgSO_4.7H_2O$ 1.2 mM, KH_2PO_4 1.2 mM, $NaHCO_3$ 25 mM, albumin 1% w/v, glucose 0.6 g/l, and 20 μM of NaF during the first 10 minutes. The stimulus of insulin secretion consisted of increasing the glucose concentrations to 3 g/l, perfusing with that buffer during the next 40 minutes. Insulin was determined in the buffer collected from the portal vein. The control group (n = 4) was perfused with a buffer with the same composition without NaF.[1]

At the end of the experiment, a perfusion with 1% w/v methylene blue was executed to visually verify the effectiveness of the perfusion.

The plasma insulin from the incubation solutions was measured with a radioimmunoassay in solid phase, using human insulin marked with ^{125}I (Diagnostic Products Corporation, Los Angeles, CA). Table 24.1 and Figure 24.1 show the results of the *in situ* perfusion of the pancreas.

The results demonstrate that both phases, initial and sustained, of the insulin secretion poststimulus with glucose are inhibited if the perfusion buffer contains 20 μM NaF. Besides, it is observed that the pancreatic tissue perfused with NaF secreted one-sixth of the insulin of the controls.

TABLE 24.1
Areas under the Curve of the Insulin Secretion in the *in situ* Perfusion of Pancreas

	Area under the Curve of the Control Animals, μIU min/ml	Area under the Curve of the Animal Treated with NaF, μIU min/ml
Initial phase	209 ± 15	107 ± 8*
Sustained phase	953 ± 67	85 ± 6*

* Represents significant differences to the controls ($p < 0.001$).

Note: The results are expressed as the mean ± SD.

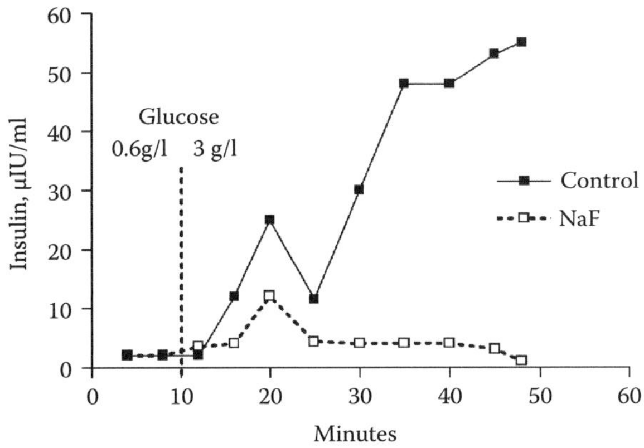

FIGURE 24.1 Insulin production by *in situ* perfused pancreas is pictured.

REFERENCE

1. Menoyo, I., A. Rigalli, and R.C. Puche. 2005. Effect of fluoride on the secretion of insulin in the rat. *Arzneimittelforschung*. 55(8): 455–460.

25 Experimental Pancreatitis

Diego Holotte, Jésica Nipoti, María J. Pretini, Verónica E. Di Loreto, and Alfredo Rigalli

EXPERIMENTAL PANCREATITIS

Utility

Incomplete closed duodenal loop (ICDL)-induced pancreatitis is a surgical technique to cause acute pancreatitis by duodenum–pancreatic reflux, which is the result of an incomplete closure of the duodenum.[1] The duodenum is ligated on half of its circumference at both sides of the duodenal entry of the biliopancreatic duct. The survival of rats with the ICDL model is higher than the complete closed duodenal loop model, and it allows for the study of the evolution from the beginning of an illness to its chronicity.

Blood flow in the pancreas and in the duodenal loop decreases over time after surgery. Histopathologic alterations of the pancreas in the ICDL group consisted of edema, parenchymal necrosis, thrombosis, and hemorrhage. From one week onward, periductal fibrosis spread to the parenchyma, with the regenerative activity of the acinar cell presenting signs of chronic inflammation.

The methodology allows the researcher to evaluate the evolution of the illness under the effect of different drugs. Several models are described that cause pancreatitis by injection of different compounds, modification of pancreatic microcirculation, or increasing the pancreobiliary duct pressure (see below). In most of the mentioned techniques, the survival of animals is not longer than 1 to 2 days.

Materials

General anesthetic
Local anesthetic
Thermostatized rat board
Ring-handled preparation scissors
Ring-handled iris scissors
Artery forceps (2)
Rat-tooth forceps
Cotton pliers (2)
Needle holder
Modeling clay
Capillaries for hematocrit
Vacuum pump
Infrared heating lamp
1-ml syringe
10-ml syringe
Injection needles 25 × 0.8 (2)
Injection needles 15 × 0.5 (2)
Iodopovidone
Antibiotic: ceftriaxone

Analgesic: diclofenac
Saline solution
Suture needle curved ½ – 5 to 10 mm (2)
Suture thread
Cotton buds (10)

Procedure

Before Surgery

Rats must be fasted for at least 12 hours, with drinking water *ad libitum*. Thirty minutes before surgery the rats are intramuscularly injected with 3 mg ceftriaxone/100 g body weight. Ceftriaxone is a cephalosporin with a broad and long action that helps to avoid postsurgery infections.

Anesthesia with ketamina/xilazine is recommended (Chapter 4). Once muscle relaxation is obtained, the rat is fixed with tape to a rat board in a dorsal decubitus position. The rat is shaved in the place where the incision will be made. The epigastric region is disinfected with iodopovidone. The absence of nociception is verified by pricking the skin between the toes with a needle or cotton pliers.

Surgery

The surgery can take from 20 to 30 minutes if there are no complications. The area of the incision is subcutaneously injected with lidocaine chlorhidrate in the line of the incision. A 1- to 2-cm median laparotomy, 5 mm in caudal direction from the xiphoid appendix, is performed. Once the stomach is visualized in the superior left region of the abdomen, the pylorus is easily identified at the lower end of the stomach. The pylorus is a sphincter that is recognized as a ring of different color from the stomach and the intestine, and it is macroscopically distinguished as a narrower zone. At the level of the pylorus, a stitch is performed with a curved needle and linen to obstruct half of the lumen of the duodenum (Figure 25.1). The knot must not be completely tightened because of the risk of necrosis. At 3 cm in the caudal direction, another ligature is performed in the same way.

After performing the incomplete duodenal close, a peritoneal wash with 10 ml of saline solution containing 10 mg/l ceftriaxone is done. The excess solution is absorbed with a vacuum pump. A piece of gauze must be placed in the end of the vacuum pump so that one does not pull viscera into

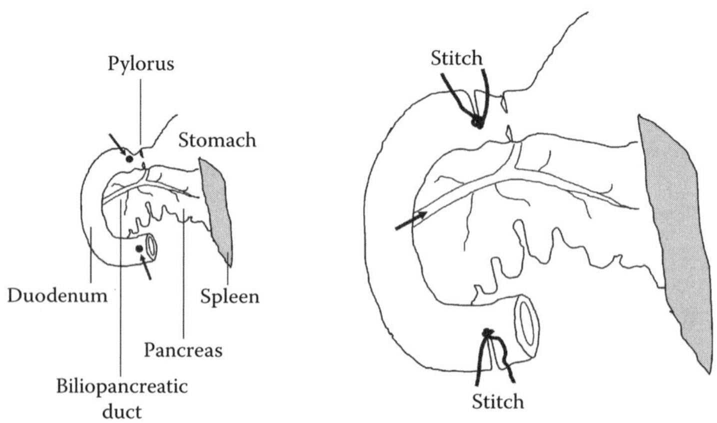

FIGURE 25.1 An incompletely closed duodenal loop model is shown. Arrows indicate the point where stitches must be done (left figure). The stitches at the level of pylorus and 3 cm in caudal direction produce an increase in the hydrostatic pressure of the duodenal content and reflux to the biliopancreatic duct (arrow, right panel).

the catheter. The abdominal muscles are sutured with 8-0 linen in a continuous suture. Finally, the skin of the animal is sutured with 5-0 nylon thread in individual stitches.

After Surgery

Immediately after surgery, 5 mg diclofenac/100 g body weight are intramuscularly injected (0.2 ml diclofenac 25 mg/ml 100 g body weight). Similar doses are injected at 8 hour intervals during the first day following surgery. In the following days, diclofenac is administered only to those rats that show clear signs of pain and suffering. The rat is placed in dorsal decubitus on a surface that is thermostatized at 30°C with the aid of an infrared heat lamp or a thermostatized rat board. The rats are maintained in this position and place until they recover mobility. Body temperature of the rats is monitored with a thermometer.

Once the rat has recovered, it is put in an individual cage, drinking water is provided *ad libitum,* and the fasted state is prolonged for another 24 hours. Food is given to the animals: 5 to 10 g of balanced food in the following 24 to 48 hours, depending on their general state. After that, food is provided *ad libitum.*

Experimental Results

The effect of the administration of disodium monofluorophosphate (MFP) on the progression of ICDL acute pancreatitis was evaluated. The study was carried out in 50-day-old male rats inbred IIM/Fm subline "m,"[2] in accordance with the international guidelines of animal care.[3]

Pancreatitis was surgically induced by the ICDL as described above.

Rats were randomly assigned to two groups: 30 days MFP-pretreated group—daily and for 30 days, 1 ml water containing 80 μmol of MFP was orally administered to a group of 21-day-old rats (treated, n = 12). Control rats received 1 ml of distilled water (n = 12). After this period, pancreatitis was induced as detailed above. In the post ICDL surgery period, general information about behavior of rats (pain, intake of food and water, amount of feces, aggressiveness, alert state) was recorded, and samples of blood were withdrawn in order to measure plasma amylase activity and alpha macroglobulin (αM) plasma levels. Amylase was measured by a kinetic method. Plasma levels of αM were measured by dot-blot with a primary anti-αM antibody raised in guinea pigs.[4]

Survival time after ICDL surgery (hours) was recorded, and data were analyzed with survival curves. Rats that died as a consequence of surgical complications were excluded in the survival curve. Those animals that died as a consequence of pancreatitis and those who survived until the end of the experiment were used for the construction of survival curves. Euthanasia was performed by intracardiac injection of 0.5 ml of saturated solution of KCl, under profound volatile anesthesia, 14 days following the surgery. After that, pancreatic and peri-pancreatic adipose tissues were removed. Tissues were fixed with 10% w/v buffered formaldehyde (pH 7), embedded in paraffin, sectioned at the 5 μm setting, and stained with hematoxylin-eosin.

All data were expressed as mean ± SEM, and analyses were done on computer software.[5] Plasma levels of amylase activities and αM concentrations were compared with repeated measures of analysis of variables (ANOVA) and Tuckey's test as posthoc comparison of means. When two sets of dependent data were analyzed, paired Student's t tests were employed. Survival curves, which plot percentage of survival rates as a function of time after surgery, were created using the method of Kaplan and Meier,[5] and calculate the 95% confidence interval (CI) for fractional survival at any particular time. Comparison of two survival curves was performed using the log-rank test. Rats that died by causes unrelated to pancreatitis were excluded from the study. In all cases, differences between data were considered significant if $p < 0.05$. Neither histopathological signs of inflammation at parenchymal pancreatic levels nor significant modifications in biochemical parameter (amylase, αM) were observed in sham operated rats (data not shown). The surgery of this group of animals (n = 6) consisted only in laparotomy, which acts as a control of the experiment.

TABLE 25.1
Forty-Eight-Hour Values of Serum Amylase and αM in Rats Treated for 30 Days with 1 ml of 80 mM MFP (MFP-treated) or the Same Volume of Water (Controls) before ICDL Surgery

	Amylase (IU/ml)			αM (μM)		
Groups	0 h	24 h	48 h	0 h	24 h	48 h
Control	29.4 ± 3.6	43.6 ± 10.0	60.4 ± 22.7	13.1 ± 1.3	9.1 ± 1.3[a]	9.8 ± 0.6[a]
MFP-treated	22.5 ± 3.2	58.7 ± 17.6	51.3 ± 14.2	14.4 ± 1.5	11.0 ± 1.6	12.0 ± 1.5

[a] Statistical differences $p < 0.05$, ANOVA, posthoc Tuckey test.
Note: The results are expressed as the mean ± SEM.

The treatment with 80 μmol MFP for 30 days before induction of pancreatitis in rats prevented death of some animals. Survival curves were significantly different in rats that received MFP from those that did not. While in control rats, 50% died within 96 hours after ICDL surgery, more than 50% of the MFP-treated rats survived after 14 days of the surgery.

Plasma activity of amylase significantly increased in both groups, with the highest value at 48 hours. Peak value of amylase was lower in MFP-treated rats, but without significant difference with respect to controls (Table 25.1). Plasma levels of αM were higher in MFP-pretreated rats at the time ICDL pancreatitis was induced. In addition, αM decreased significantly in controls, while in MFP-pretreated animals it always remained above values of controls (Table 25.1).

Histopathologic alterations of the pancreas in the ICDL group consisted of parenchymal necrosis, fibrin deposition, and leukocite migration with polymorphonuclear cell infiltration.

Chronic scarring, inflammatory changes, and some areas of preserved parenchyma have been observed microscopically in MFP-pretreated animals. The presence of hemosiderophages indicates a previous hemorrhagic episode. A few animals did not show the presence of important histological alteration, which has been interpreted as a *restitutio ad integrum* (full repair or recovery) of the pancreatic tissue.

Other Models of Experimental Pancreatitis

1. *Necrotizing and hemorragic acute pancreatitis*
   It can be induced by intraperitoneal injection of 40 μg cerulein/kg plus bradykinin. Administration of bradykinin reduces amylase levels without changes in histology.[6] Another model of this type of pancreatitis can be induced by sodium taurocholate: retrograde infusion of 0.2 ml of 2.5% w/v sodium taurocholate in the cannulated pancreatobiliary duct. Serum amylase and lipase increase in 5 hours, severity of tissue damage is not correlated with enzyme levels. Histological changes are present in 6 hours. Similar results are found with 0.2 ml glycodeoxycholate 10 mM or the mixture of 0.2 ml of glycodeoxycholate and 10 U enterokinase. After 14 days, it results in widespread chronic inflammation, acinar dilatation atrophy, marked reactive stromal proliferation, and/or ductular budding with periductal fibrosis.[7] Necrotizing and hemorrhagic pancreatitis can also be induced by trypsin, which is injected into the interlobular tissue of the duodenal part of the pancreas. After 6 hours, amylase increases and edema is detected.
2. *Edematous pancreatitis*
   Subcutaneous infusion of cerulein for 5 hours (10 μg/kg/h) produces edema, leukocyte infiltration, and vacuolization of acinar cells after 12 to 18 hours. After 14 days, histopathologic features are not different from controls.[8] There is a simplified method by subcutaneous injection of 20 μg/kg and 20 μg/kg intravenously one hour later.[9]

3. *Acute pancreatitis by excessive dose of caerulein*
   Rats are infused via a jugular vein catheter with 5 μg/kg/h for 24 hours. Edema is seen after 1 hour and plasma amylase activity reaches a 10-fold elevation.[10]
4. *Ischemic acute pancreatitis*
   It is produced by clamping of the supplying artery for 40 minutes after obstruction of the biliopancreatic duct for 20 minutes or hyperstimulation with 5 U intravenous injection of cholecystokinin. The treatment produces fat necrosis and/or acinar necrosis at the peripheral lobules.[11] Ischemic pancreatitis can also be produced by injection of polystyrene microspheres of 20-μm diameter into the pancreatic branches of the splenic artery. It develops hemorrhagic pancreatitis by 27 hours following surgery. Serum amylase increases twofold. It produces features of chronic pancreatitis: intesticial fibrosis, ductal changes, mononuclear infiltration, acinar necrosis, and atrophy.[12]
5. *Acute pancreatitis and normal tissue in the same rat*
   Ligation of the main pancreatic duct of the splenic lobe and intact maintenance of the gastroduodenal lobe in the same rat. This model produces damaged pancreatic tissue and normal tissue in the same animal.[13]
6. *Edematous and necrotizing pancreatitis*
   It can be induced by closed duodenal loop (CDL), which produces a model of edematous and necrotizing pancreatitis in approximately 6 hours that finally produces the death of the animal.[14] Edematous and necrotizing acute pancreatitis can also be induced by injection of phospholipase A2 300 ng/kg. It produces intersticial edema in the pancreas and lung, and neutrophil infiltration with acinar necrosis in the pancreas. Phospholipase A2 activity is not necessary for induction of acute pancreatitis.[15] Necrotizing and edematous acute pancreatitis can be induced by hypercalcemia by a bolus infusion of $CaCl_2$ 200 mg/kg. One hour later, plasma amylase activity increases two and a half fold and trypsinogen activation peptide threefold. Edema and leukocyte infiltration are observed. Amylase and trypsinogen activation peptide remain high after 24 hours and necrosis becomes evident.[16]
7. *Edematous with focal necrosis pancreatitis*
   Acute pancreatitis is induced by oxygen radicals, which are produced by infusion in the biliopancreatic duct of $FeSO_4$ or H_2O_2 or both.[17] Edematous acute pancreatitis can also be obtained by induction with ethanol. Four hours of intravenous infusion of 0.5 to 1 g ethanol/kg/h caused mild but significant rises in serum amylase, pancreatic water content, and intersticial edema.[18]
8. *Necrotizing acute pancreatitis*
   This kind of acute pancreatitis can be induced by the intraperitoneal injection of 500 mg/100 g body weight of L-arginine. Damage appears after 6 hours, with edema at 24 hours, and focal cellular necrosis at 48 hours, with a maximum severity after 72 hours. Amylase rises at 12 to 24 hours, decreases below controls at 72 hours to 7 days, and returns to normal values after two weeks. The content of pancreatic proteins, DNA, and digestive enzymes are markedly reduced after the arginine injection and reach their nadirs at 72 hours.[19] Another way of producing necrotizing pancreatitis is by ischemia/reperfusion. It is a model of necrotizing pancreatitis with pancreatic regeneration. Blood flow in the inferior splenic artery is limited for 30 minutes, followed by reperfusion. Amylase increases between 12 hours and 3 days. Pancreatic tissue shows necrosis, edema, and leukocyte infiltration. Maximal damage is observed between 1 and 2 days of reperfusion. Between the second and tenth day, signs of chronic pancreatitis, such as fibrosis, acinar cell loss, formation of tubular complexes, and dilatation of ducts are observed.[20] Mild necrotizing acute pancreatitis can be obtained by infusion of very low concentrations of glicodeoxicholate with intravenous cerulein. After 14 days, moderate degrees of chronic inflammation and acinar atrophy, along with an intermediate degree of periductal fibrosis and stromal reaction, are observed.[21]

9. *Diffuse necrosis of pancreas and fat*
   It is obtained by intraparenchymal pancreatic injection of 0.3 ml/100 body weight of 5% or 10% w/v taurocholate. Amylase and lipase sharply increase after 6 hours and normalize at 24 hours. It is a reproducible model and the damage, which is dose dependent, correlates with histopathological examination and mortality.[22] Ligation/obstruction of the biliopancreatic duct produces changes similar to that in human beings.[23]
10. *Hemolysis-induced acute pancreatitis*
    Hemolysis is induced by intraperitoneal injection of 60 mg/kg of 20 mg/ml acetylphenylhydrazine in 20% v/v ethanol. Rats with massive hemolysis had hyperamylasemia. Severe pancreatitis occurs on day 3.[24]
11. *Reversible pancreatitis*
    Common bile duct is occluded for 24 hours and, after that, the blocker is eliminated. Edema, zymogen degranulation, inflammatory infiltration, vacuolization of acinar cells, focal areas of fat, and parenchymal necrosis are present after 1 week. After 3 weeks, it shows normal appearance.[25]
12. *Hemorrhagic acute pancreatitis*
    It can be induced by local retrograde infusion into the splenic artery of divalent cations, such as Ca^{2+}, Mn^{2+}, Zn^{2+}, or Ba^{2+}, 0.6 mmol/kg/h for 3 hours at 0.1 ml/h.[26]
13. *Choline deficient ethionine supplemented diet model pancreatitis in mice.*[27]
14. *Models of pancreatic fibrogenesis*
    14.1 *Severe hyperstimulation and obstructive pancreatitis*
    Complete pancreatic duct obstruction and daily cerulein intraperitoneal (50 μg/kg) injection reproduce early stages of chronic pancreatitis. They produce edema and inflammation, but minimal paranchymal necrosis. After 48 hours, there is significant loss of acinar mass, and then normal acinar cells are absent after 96 hours. Large amounts of collagen were found after 72 hours.[28]
    14.2 *Serial injection of* L*-arginine*
    Intraperitoneal injection of 500 mg/100 g body weight is followed by three injections of 250 mg over 100 days. Amylase increases only in the acute phase of edematous pancreatitis. By day 5, there is up to 90% acinar destruction, and adipose replacement. Ductal, vascular, and islet cells appear undamaged. Changes are present 6 months after injections.[29]
15. *Autoimmune pancreatitis*
    A recently defined disease of unknown etiology, autoimmune pancreatitis is characterized by inflammatory infiltrates in the pancreas with conspicuous involvement of the ducts. The disease clinically manifests in humans by epigastric pain, weight loss, and jaundice. It can be produced in rats by transference of amylase-specific $CD4^+$ sensitive T cells to a syngeneic recipient animal. Clinically, it is similar to humans.[30]

REFERENCES

1. Sugimoto, M., T. Takada, and H. Yasuda. 2004. A new experimental pancreatitis by incomplete closed duodenal loop: The influence of pancreatic microcirculation on the development and progression of induced severe pancreatitis in rats. *Pancreas* 28(4): e112–119.
2. Calderari, S., M.T. Font, O. Garrocq, S.M. Martínez, R.C. Puche, M.C. Tarrés. The inbred IIM/FcM stock. *Rat Newsl.* 25: 28–29.
3. U.S. Department of Health and Human Services. NIH Publication No. 86-23, revised 1985, National Institutes of Health, Bethesda, MD.
4. Hoffmann, E.M., S. Muetzel, and K. Becker. 2002. A modified dot-blot method of protein determination applied in the tannin-protein precipitation assay to facilitate the evaluation of tannin activity in animal feeds. *Br. J. Nutr.* 87:.421–426.

5. Kaplan, E., and P. Meier. 1994. Graph Pad Prism [computer program], v. 2.0. GraphPad Software, San Diego, CA.
6. Weidenbach, H., M.M. Lerch, M.H. Schoenberg, T.M. Gress, S. Turi, and G. Adler. 1996. Characterization of a non-invasive, vascular model of acute necrotizing pancreatitis. *Z. Gastroenterol.* 34(1): 9–14.
7. Conde, J.G., J.M. Lopez Novoa, C. Novo Alonso, and J.M. Romeo Martinez. 1990. A model of experimental acute pancreatitis in the conscious rat. *Rev. Esp. Enferm.* Dig. 77(6): 433–436.
8. Jaworek, J., J. Bonior, K. Nawrot, A. Leja, R. Sendur, J. Stachura, W. Pawlik, and S. Konturek. 2002. Intracerebroventricular administration of bacterial lipopolysaccharide prevents the development of acute experimental pancreatitis in the rat. *Med. Sci. Mon.* 4: BR136–143.
9. Abdo, E.E., A.M. Coelho, A.L. Montagnini, M.S. Kubrusly, K.R. Leite, S.N. Sampietre, N.A. Molan, M.C. Machado, and H.W. Pinotti. 1994. Simplified model of induction of experimental acute pancreatitis with a supra-maximal dose of cerulein. *Rev. Hosp. Clin. Fac. Med. Sao Paulo* 49(5):.204–207.
10. Lampel, M., and H.F. Kern. 1977. Acute interstitial pancreatitis in the rat induced by excessive doses of a pancreatic secretagogue. *Virchows Arch. A. Pathol. Anat. Histol.* 373(2): 97–117.
11. Siech, M., H. Weber, G. Letko, W. Dummler, M.H. Schoenberg, and H.G. Beger. 1997. Similar morphological and intracellular biochemical changes in alcoholic acute pancreatitis and ischemic acute pancreatitis in rats. *Pancreas* 14(1): 32–38.
12. Redha, F., G. Uhlschmid, R.W. Ammann, and A.U. Freiburghaus. 1990. Injection of microspheres into pancreatic arteries causes acute hemorrhagic pancreatitis in the rat: A new animal model. *Pancreas* 5(2): 188–193.
13. Satake, K., and A. Hiura. 1998. A new model for pancreatitis. *Pancreas* 16(3): 284–288.
14. Nevalainen, T.J., and A. Seppa. 1975. Acute pancreatitis caused by closed duodenal loop in the rat. *Scand. J. Gastroenterol.* 10(5): 521–527.
15. Camargo, E.A., L.C. Esquisatto, M.A. Esquisatto, M.T. Ribela, A.C. Cintra, J.R. Giglio, E. Antunes, and E.C. Landucci. 2005. Characterization of the acute pancreatitis induced by secretory phospholipases A(2) in rats. *Toxicon*: *Off. J. Int. Soc. Toxinol.* 46(8): 921–926.
16. Mithofer, K., C. Fernandez-del Castillo, T.W. Frick, K.B. Lewandrowski, D.W. Rattner, and A.L. Warshaw. 1995. Acute hypercalcemia causes acute pancreatitis and ectopic trypsinogen activation in the rat. *Gastroenterology* 109(1): 239–246.
17. Coskun, T., P. Korkusuz, Y. Kaya, U. Ors, Y. Aker, and K. Kilinc. 2003. Free oxygen radical-induced acute pancreatitis. A light and electron microscopic study. *Hepatogastroenterology* 50(49): 43–48.
18. Pandol, S.J., I. Gukovsky, A. Satoh, A. Lugea, and A.S. Gukovskaya. 2003. Animal and *in vitro* models of alcoholic pancreatitis: role of cholecystokinin. *Pancreas* 27(4): 297–300.
19. Tani, S., H. Itoh, Y. Okabayashi, T. Nakamura, M. Fujii, T. Fujisawa, M. Koide, and M. Otsuki. 1990. New model of acute necrotizing pancreatitis induced by excessive doses of arginine in rats. *Dig. Dis. Sci.* 35(3): 367–374.
20. Hoffmann, T.F., R. Leiderer, H. Waldner, S. Arbogast, and K. Messmer. 1995. Ischemia reperfusion of the pancreas: A new *in vivo* model for acute pancreatitis in rats. *Res. Exp. Med.* (Berlin) 195(3): 125–144.
21. Schmidt, J., C.C. Compton, D.W. Rattner, K. Lewandrowski, and A.L. Warshaw. 1995. Late histopathologic changes and healing in an improved rodent model of acute necrotizing pancreatitis. *Digestion* 56(3): 246–252.
22. Paran, H., A. Mayo, D. Kidron, G. Sivak, T. Reshef, T. Vider, O. Ziv, and U. Freund. 2000. Experimental acute necrotising pancreatitis: Evaluation and characterisation of a model of intraparenchymal injection of sodium taurocholate in rats. *Eur. J. Surg.* 166(11): 894–898.
23. Urunuela, A., M.A. Manso, A.M. de la Mano, and I. de Dios. 2003. Cholecystokinin blockade triggers earlier and enhanced intra-acinar oxygen free radical generation in acute pancreatitis induced by pancreatic duct obstruction in rats. *Clin. Sci.* (London) 105(2): 203–212.
24. Saruc, M., H. Yuceyar, N. Turkel, O. Ozutemiz, I. Tuzcuoglu, G. Yuce, and A. Huseyinov. 2003. An experimental model of hemolysis-induced acute pancreatitis. *Braz. J. Med. Biol. Res.* 36(7):.879–386.
25. Azima, B., R.L. Kao, G. Youngberg, D. Williams, and W. Browder. 1996. A new animal model of reversible acute pancreatitis. *J. Surg. Res.* 63(2): 419–424.
26. Kaiser, A., T. Frick, S. Scharen, A. Freiburghaus, and F. Largiader. 1993. Acute pancreatitis after local infusion of divalent cations. *Helv. Chir. Acta.* 59(5-6): 897–960.
27. Rakonczay, Z. Jr., T. Takacs, B. Ivanyi, Y. Mandi, G. Papai, I. Boros, I. Varga, K. Jost, and J. Lonovics. 2002. Induction of heat shock proteins fails to produce protection against trypsin-induced acute pancreatitis in rats. *Clin. Exp. Med.* 2(2):.89–97.

28. Murayama, K.M., B.L. Barent, M. Gruber, A. Brooks, S. Eliason, E.M. Brunt, and G.S. Smith. 1999. Characterization of a novel model of pancreatic fibrosis and acinar atrophy. *J. Gastrointest. Surg.* 3(4): 418–25.
29. Delaney, C.P., K.F. McGeeney, P. Dervan, and J.M. Fitzpatrick. 1993. Pancreatic atrophy: A new model using serial intra-peritoneal injections of L-arginine. *Scand. J. Gastroenterol.* 28(12):.1086–1090.
30. Davidson, T.S., D.S. Longnecker, and W.F. Hickey. 2005. An experimental model of autoimmune pancreatitis in the rat. *J. Pathol.* 166(3): 729–736.

Section VI

Liver

26 Hepatic Circulation Impairment

Gabriel A. Inchauspe and Alfredo Rigalli

HEPATIC CIRCULATION IMPAIRMENT

UTILITY

This technique is useful when the plasma clearance of a substance by the liver is investigated. Information about whether the liver is involved in the plasma clearance or not can be obtained with this technique. Samples of blood are obtained through a venous or arterial catheter before and after the hepatic circulation is impaired, which is described below. In the samples obtained before the maneuver, the liver is included in the clearance, and in the samples obtained after the impairment of circulation, the clearance of the substance is from all tissues except the liver.

Materials

General anesthetic
Thermostatized rat board
Ring-handled preparation scissors
Ring-handled iris scissors
Cotton pliers (2)
Artery forceps (6)
Rat-tooth forceps
Needle holder
Saline solution
Iodopovidone
Suture needle curved ½ – 5 to 10 mm
Suture thread
Cotton
Cotton buds (10)
Gauze

These materials are necessary for the hepatic circulation impairment *in situ*. In case the technique is combined with other techniques, e.g., arterial catheterization, tracheostomy, etc., the specific materials for these techniques must be reconsidered.

PROCEDURE

After general anesthesia is administered, the femoral artery is cannulated to obtain blood samples. Five minutes before hepatic circulation impairment, a 3-cm median laparotomy is performed. The gut is moved to the left with the aid of cotton buds in order to visualize the celiac artery and the portal vein. A thread is passed under both vessels with the aid of two cotton pliers or a curved needle. Knots are made, but they are not fixed. The threads passed under both vessels can be identified by clamping them with different artery forceps. When hepatic circulation needs to be stopped, both knots are fixed. The excess of thread is cut and the abdominal incision is sutured. The described maneuver can take as little as 5 minutes for a skilled surgeon. In this way, dehydration of the rats is minimal and the survival rate of the animal is greater.

Experimental Results

Two-hundred-gram female inbred IIM/Fm rats were used to study the effect of growth hormones on plasma levels of triacylglycerols. The femoral artery was cannulated and blood samples were obtained at 0, 10, 20, 35, 50, 65, 80, 95, 110, and 125 minutes. At the beginning of the experiment (0 min), 0.5 ml of heparin was injected. After 20 minutes, 2 μg of human recombinant growth hormone (hrGH), dissolved in 0.5 ml of saline solution, were injected. Sixty-five minutes from the beginning of the experiment, hepatic circulation impairment was performed as mentioned above.

At different times during the study (Figure 26.1), 100 to 200 μl of blood were obtained through the catheter inserted in the femoral artery. Blood was centrifuged and plasma was saved for triacylglycerols and glucose measurement. The liquid from the bottom of the centrifuge tube, containing blood cells, was diluted with an equal volume of saline solution and reinjected into the rat through the same catheter. Although this procedure causes a small (less than 1%) hemodilution, it improves the survival of animals.

Plasma levels of triacylglycerols linearly increased as a function of time in rats that received hrGH (see Figure 26.1). In controls without hrGH, the slope of the linear regression remained at zero.

Glucose displayed a predictable variation throughout the experiment in both groups of rats (Figure 26.2). At the beginning of the experiment, plasma glucose increased as a consequence of stress in the animal. After the hrGH injection, glucose increased as a consequence of its hyperglycemic effect. The impairment of hepatic circulation caused, in both groups, a linear decrease in plasma levels of glucose that did not differ in slope.

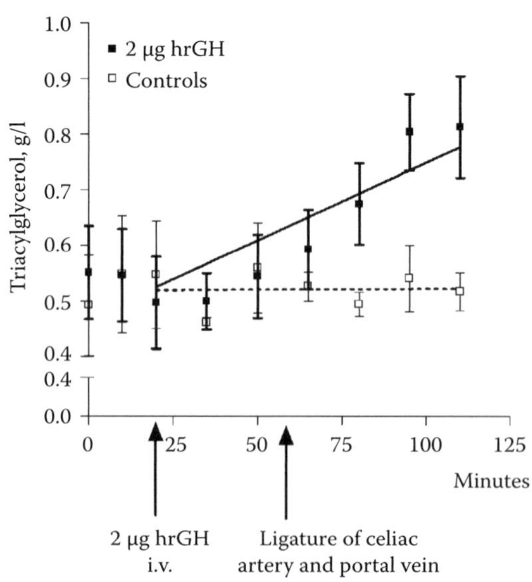

FIGURE 26.1 Shown are plasma levels of triacylglycerols in rats after the injection of 0.5 ml of saline solution with or without 2 μg of hrGH.

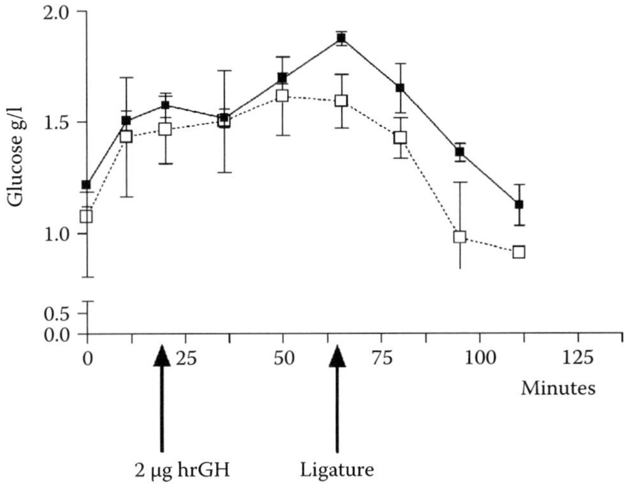

FIGURE 26.2 Shown are plasma levels of glucose in rats after the injection of 0.5 ml of saline solution with or without 2 μg of hrGH.

27 Extrahepatic Cholestasis Model

Anabel Brandoni and Adriana M. Torres

INTRODUCTION

Cholestasis is characterized by the retention of bile and bile components. Chief clinical manifestations of cholestasis include jaundice, pruritus, and elevations of bile salts, bilirubin, alkaline phosphatase, and related enzymes in serum.

On a mechanistic basis, cholestasis usually is divided into "extrahepatic" and "intrahepatic" forms. The first refers to obstruction of large bile ducts outside the liver, e.g., due to gallstones, while the causes of intrahepatic cholestasis lie within the liver. Extrahepatic mechanical cholestasis is not a rare abnormality. It occurs in about 10% of patients suffering from cholelithiasis and in the majority of neoplasms affecting the pancreas and the common bile duct.[1]

UTILITY

Experimental models that simulate human cholestasis have been very useful in furthering the understanding of its pathogenesis. Bile duct ligation is the most frequent model for extrahepatic mechanical obstruction in humans. Studies of obstructive jaundice have been widely conducted in animal models, and the rat has become the most widely used experimental animal for research of obstructive jaundice.

Materials

General anesthetic
Thermostatized rat board
Ring-handled preparation scissors
Ring-handled iris scissors
Cotton pliers
Adson forceps
Artery forceps
Rat tooth forceps
Needle holder
Spatula
Catheter
Injection needle 15 × 0.5
Saline solution
Heparin
Iodopovidone
Suture needle ½ – 5 to 10 mm
Suture thread
Cotton buds

Procedure

Obstructive jaundice is induced in rats by double ligation and the division of the common bile duct. In order to perform these surgical procedures, animals are anesthetized and placed in a sterile operating field. The upper middle part of the abdomen is washed with saline solution. The abdomen is shaved with an electrical surgical shaver while the hair is still wet. This makes shaving easier and prevents the generation of airborne debris and hair. A 1.5- to 2.0-cm upper midline abdominal incision is made. A section of duodenum is selected at approximately 1.5 cm from the pyloric sphincter. Then the duodenum is pulled out of the abdominal cavity, grasped gently, and displaced anteroinferiorly to place light tension on the biliary tract. Following this, the common bile duct is isolated using very fine forceps. Then two cotton threads are passed under the bile duct, guided by the forceps. Finally, obstructive jaundice is induced by tying the ligatures close to the liver hilus immediately below the bifurcation, and dividing the common bile duct between the ligatures (Figure 27.1). The abdominal wall is closed in two layers by single sutures. The entire procedure takes 10 to 15 min to complete.

All animals are allowed free access to a standard laboratory food and tap water, and housed in a constant temperature and humidity environment with regular light cycles (12 hours) during the experiment. Studies can be performed at different times after surgery.

Evaluation of Experimental Rat Model of Cholestasis

Biochemical Determinations

The day of the experiment, blood samples are withdrawn and used to measure total and direct bilirubin as indicative parameters of hepatic function. These biochemical analyses are performed using a commercial kit (Wiener Lab, Rosario, Argentina).

Pathology

Tissue samples of the liver are taken when the rats are sacrificed at the end of the experiment. Morphology of these tissues is examined with routine hematoxylin and eosin staining and also with hematoxilin and periodic acid-Schiff (PAS).

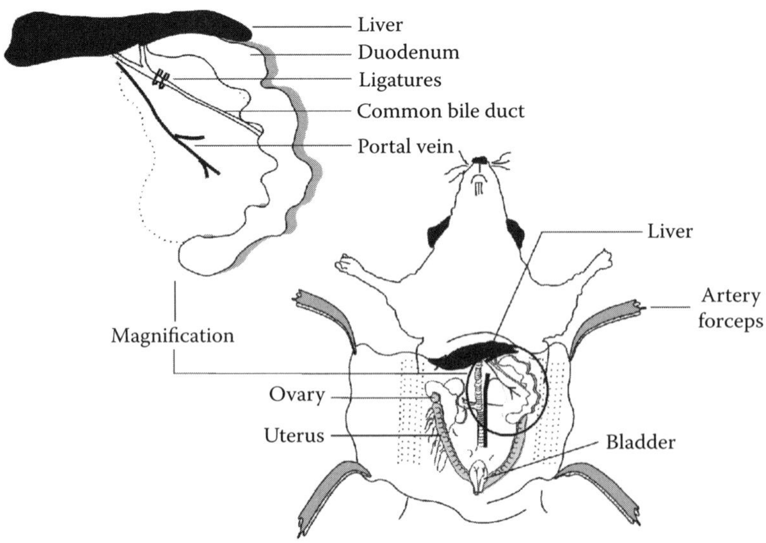

FIGURE 27.1 Area of ligature of the common bile duct is pictured.

TABLE 27.1
Total and Direct Bilirubin Plasma Levels in Sham and BDL Rats

	21 h Postsurgery		72 h Postsurgery	
	Sham (n = 5)	BDL (n = 6)	Sham (n = 4)	BDL (n = 5)
Total bilirubin (mg/l)	5.3 ± 0.5	43.7 ± 3.0 *	2.5 ± 0.4	71.3 ± 4.3 *
Direct bilirubin (mg/l)	2.3 ± 0.2	34.4 ± 3.7 *	1.3 ± 0.2	66.8 ± 4.2 *

* $p < 0.05$ compared with the respective Sham group.
Note: The results are expressed as the mean ± SEM.

Experimental Results

Male Wistar rats aged from 110 to 130 days were used throughout the study. Animals were anesthetized and bile duct ligation was performed. Studies were carried out 21 and 72 hours after surgery. Controls underwent sham operations that consisted of exposure, but not ligation, of the common bile duct (Sham group).

On the day of the experiment, animals were anesthetized, blood was withdrawn from the femoral artery through a catheter previously inserted for both the Sham and bile duct-ligated group (BDL). Plasma was separated by centrifugation, and the samples were used to measure total and direct bilirubin plasma levels.

Bile duct ligation in rats ended in severe cholestasis, as indicated by increases of plasma levels of total and direct bilirubin (Table 27.1).[2-5] These findings corroborate the adequacy of the study design.

Animals were sacrificed at the end of the study. The liver and kidneys were removed for further histological examination under light microscopy. The structure of the liver from the control group (Sham) was of normal morphology, whereas in animals with bile duct obstruction, several abnormalities were detected. An enlargement of portal spaces related to interstitial edema was seen. Some mononuclear cell infiltrations were noted around portal triads. Numerous and dispersed areas of recent necrosis were also detected.

PAS staining revealed the normal disposition of the liver in the Sham rats. In animals with mechanical cholestasis, an evidence in diminished PAS staining was observed. This observation is related to the presence of recent necrosis areas.[5]

REFERENCES

1. Reichen, J., and F.R. Simon. 1988. Cholestasis, In *The liver: Biology and pathobiology*, 2nd ed., I.M. Arias et al., Eds. New York: Raven Press, pp. 1105–1124.
2. Brandoni, A., N.B. Quaglia, and A.M. Torres. 2003. Compensation increase in organic anion excretion in rats with acute biliary obstruction: Role of the renal organic anion transporter 1. *Pharmacology* 68: 57–63.
3. Brandoni, A., S.R. Villar, J.C. Picena, N. Anzai, H. Endou, and A.M. Torres. 2006. Expression of rat renal cortical OAT1 and OAT3 in response to acute biliary obstruction. *Hepatology* 43: 1092–1100.
4. Brandoni, A., N. Anzai, Y. Kanai, H. Endou, and A.M. Torres. 2006. Renal elimination of p-aminohippurate (PAH) in response to three days of biliary obstruction in the rat. The role of OAT1 and OAT3. *Biochim. Biophys. Acta.* 1762: 673–682.
5. Brandoni, A. 2006. Depuración renal y sistémica de aniones orgánicos en ratas con colestasis extrahepática. Doctoral thesis, Facultad de Ciencias Bioquímicas y Farmacèuticas, Universidad Nacional de Rosario, Rosario, Argentina.

Section VII

Ablation of Endocrine Glands

28 Adrenalectomy

Laura I. Pera and Alfredo Rigalli

INTRODUCTION

The adrenal glands are located in the retroperitoneal fat pad on both sides of the spine, in the upper end of each kidney, slightly moved toward the inner side of each kidney (Figure 28.1). Once excised, they are brownish structures 0.2-cm thick, 0.5-cm long, and 0.3-cm wide, and resemble grape seeds. This aspect is observed after the excision of the gland. *In situ*, they are clear, brownish, rounded 0.5-cm-long bodies in the fat tissue. They present an anterior, an outer, and an inner side, and they are protected by a capsule that exhibits a soft and brilliant aspect.

The age of the rat influences the location of the adrenal glands. After weaning, adrenal glands are located near the upper end of the kidney. As rats grow, glands separate from the kidney and are included in the fat tissue pad, over the kidney, which makes them harder to identify and extract. Also, in adult rats, adrenal glands are smaller than in young rats.

Adrenal glands consist of an internal medulla and an external cortex. The cortex secretes three kinds of hormones:

1. *Glucocorticoids*: cortisol and cortisone. They have antiinflammatory effects and induce changes in carbohydrate and lipid metabolism.
2. *Mineralocorticoids*: aldosterone. It plays a key role in the hydroelectrolitic balance of sodium and potassium.
3. *Sexual steroids*: androgens and estrogens. They contribute to the development of secondary sexual characters.

Hormones secreted by the cortex of the adrenal glands are synthesized from cholesterol by chemical modification of cyclopentanoperhydrophenanthrene. Cholesterol is actively taken up from blood by specific receptors for low-density lipoproteins (LDLs) of cells in the cortex or synthesized in the cells from acetyl-CoA. Cortex adrenal hormones help the tissue to achieve high rates of metabolic processes needed to maintain homeostasis.

The rat with an adrenalectomy is a model to study the effect of adrenal hormones or the effects of treatments in the absence of those hormones. Adrenalectomy produces a decrease in plasma aldosterone, with a subsequent decrease in plasma sodium; as a consequence, adrenalectomized rats must receive NaCl 9 g/l instead of drinking water. On the other hand, 15 g/l glucose must be administered as drinking water because of a paucity of glucocorticoids that predisposes to hypoglycemia. Furthermore, adrenalectomy produces decrease in weight gain because of lower food intake than control animals. In this sense, a sham-operated fed rat pair must be used as controls.

The supplementation with hydrocortisone corrects the weight gain. On the other hand, the administration of aldosterone corrects the plasma levels of sodium and potassium.

When the effects of mineralocorticoids are studied, hydrocortisone 0.5 mg/day 100 g body weight or dexamethasone 1.2 μg/day 100 g body weight must be administered; whereas, if the effects of glucocorticoids are studied, aldosterone (0.5 μg/day 100 g body weight) should be administered to control plasma concentrations of sodium and potassium. The administration of hormones is preferably done with an osmotic minipump that produces a constant concentration of the hormone.

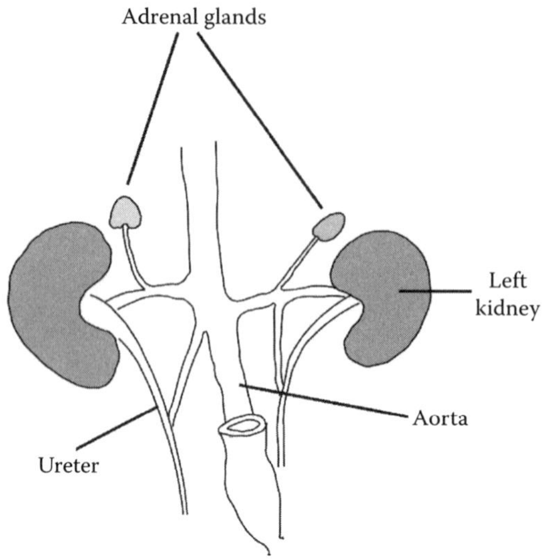

FIGURE 28.1 Shown is the location of adrenal glands. Gut has been removed in order to make clear the relationship between the glands and the kidneys.

Materials

General anesthetic
Thermostatized rat board
Ring-handled preparation scissors
Ring-handled iris scissors
Cotton pliers (2)
Adson forceps
Artery forceps (2)
Rat-tooth forceps
Needle holder
Iodopovidone
Suture needle ½ – 5 to 10 mm
Suture thread
Cotton
Cotton buds (10)

Procedure

After anesthesia, the rat is held in the rat board in dorsal decubitus, and a median supraumbilical 3-cm laparotomy is made. Two artery forceps are used to obtain a wide surgical area. As mentioned above, glands are located above the kidney, near the upper end, in the adipose tissue. Adrenal glands are irrigated by an artery branch of the renal artery. In young animals, they are easily distinguished and must be clamped with a 10/0 linen thread. In some cases, if the vascular pedicle is not visible, glands can be excised without clamping them because usually the hemorrhage that the ablation produces is negligible. Then, the gland is held with an Adson forceps and it is separated from the adipose tissue with cotton pliers while pulling with the Adson forceps. Ablation of the right gland is needed to move the gut to the left and must be performed carefully, in order to not damage the liver. On the other hand, ablation of the left gland is needed to move the

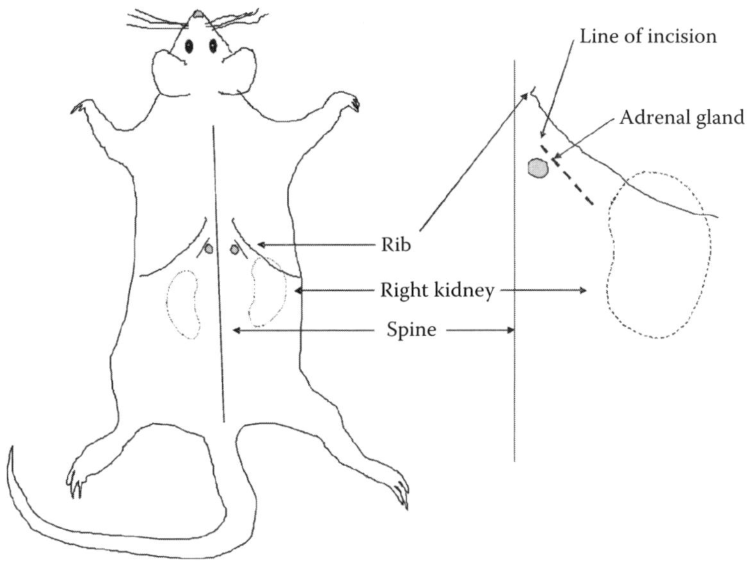

FIGURE 28.2 Lines of lumbar incisions and location of adrenal glands with respect to kidneys, spine, and ribs are pictured.

gut to the right and great care must be taken with the spleen, which is over the left kidney and the adrenal gland.

Ablation of adrenal glands can also be performed through two 1-cm lumbar incisions at both sides of the spine (Figure 28.2). Adrenal glands are easily visible in the adipose tissue immediately after the incision of the skin and muscles.

Experimental Results

Plasma levels of sodium and potassium were measured in intact and in adrenalectomized rats with and without hormone replacement (Table 28.1). Twenty-five male 9-week-old Sprague Dawley animals were bilaterally adrenalectomized (Adx) 7 days before the study. Rats were randomly divided into 5 groups, and an osmotic minipump was subcutaneously inserted with vehicle (polyethylene glycol) or vehicle with aldosterone (A) and/or dexamethasone (D):

TABLE 28.1
Plasma Levels of Aldosterone, Sodium, and Potassium after 7 Days of Adrenalectomy

Groups	Aldosterone, ng/dl	Sodium meq/l	Potassium meq/l
Controls	7.2 ± 1.3	145.7 ± 1.0	4.45 ± 0.10
Adx	0.5 ± 0.3*	138.8 ± 0.8*	5.51 ± 0.14*
Adx +A	4.4 ± 1.9	144.3 ± 0.8	4.59 ± 0.15
Adx +D	0.5 ± 0.2*	139.6 ± 1.0*	4.92 ± 0.15*
Adx + A + D	4.6 ± 1.9	147.9 ± 1.0	4.43 ± 0.12

* Represents significant differences with respect to controls $p < 0.05$.
Note: Data are expressed as mean ± SEM.

- Controls: Sham operated rats + vehicle
- Adx: adrenalectomized rats + vehicle
- Adx + A: adrenalectomized rats + 0.5 μg aldosterone/day 100 g body weight
- Adx + D: adrenalectomized rats + 1.2 μg dexamethasone/day 100 g body weight
- Adx + A + D: adrenalectomized rats + 0.5 μg aldosterone/day 100 g body weight + 1.2 μg dexamethasone/day 100 g body weight.

Rats were maintained on an 18-g/day standard rat chow. As for drinking water, controls and Adx + A + D rats received tap water, Adx rats received tap water containing NaCl 9 g/l and glucose 15 g/l, Adx + A: tap water containing 15 g/l glucose, and Adx + D: tap water containing NaCl 9 g/l.

29 Ovariectomy

Verónica E. Di Loreto, Laura I. Pera, and Alfredo Rigalli

OVARIECTOMY

Utility

Ovariectomy (OVX) is applicable in experiments in which some effects related to ovarian hormone deficiency can be studied. In this procedure, the ablation of both ovaries is performed.

In women, a reduction of the estrogen levels and the loss of the reproductive capacity are observed at the beginning of menopause. This stage in life goes together with numerous changes in the genitourinary and cardiovascular system, as well as in skin, hair, and bones. Estrogen decrease produces an increase of the bone turnover, which means an increase in the resorption and formation of bone tissue. Resorption is faster than formation so that a loss of trabecular bone occurs, causing a subsequent loss of bone strength. The ovariectomized rat is a model applicable to studies concerning trabecular bone loss caused by induced menopause. The characteristics of this model have been completely described,[1,2] and it is frequently used as a model for postmenopausal osteoporosis. However, the rat does not develop fractures, the hallmark of osteoporosis. It is applicable mainly to the study of situations in which the bone fragility increases through resorption, but not to those in which bone formation decreases, as happens with immobility.

Ovariectomy is a regulatory exigence to assess the efficacy and safety of new agents for osteoporosis therapy. The FDA Guidelines[3] and WHO Guidelines[4] recommend that agents must be evaluated in two animal species, including the OVX rat and a second nonrodent model (e.g., sow, ewe, primate, etc.).

Materials

General anesthetic
Thermostatized rat board
Ring-handled preparation scissors
Cotton pliers (2)
Artery forceps (2)
Needle holder
Iodopovidone
Suture needle curved ½ – 5 to 10 mm
Suture thread
Cotton buds (5)

Procedure

The rat must be anaesthetized and placed in a supine position. There are two possibilities in performing the surgery: a longitudinal median laparotomy of 2 to 3 cm in the lower area of the abdomen or two paramedian ventral incisions of 1 cm each. The ovaries are small organs of 0.5 cm, with a pink-red irregular surface, and they are connected to each oviduct, which connect to the uterus (Figure 29.1). The uterus is a pink, long duct with a characteristic irrigation. The uterus must be

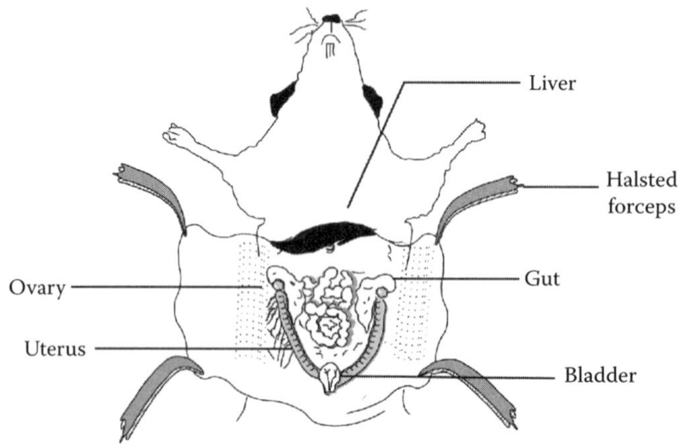

FIGURE 29.1 Location of ovaries and uterus is pictured.

located at one side of the bladder and once found, must be followed in an upward direction until the ovary is found.

Two bonds must be made, one between the ovary and the uterus and the other one in the superior end, bonding the ovarian artery (Figure 29.2). Then, the ovary is extracted using ring-handled preparation scissors and cotton pliers.

Characteristics of the Rat Ovariectomized Model

The OVX produces an imbalance in bone remodeling with a greater increase of bone resorption rather than bone formation. Osteopenia is more evident in trabecular areas of metaphysis of long bones.[5] The model produces an increase of the resorption markers (N-telopeptides, C-telopeptides, hydroxyproline, pyridinolines) as well as in formation markers (alkaline phosphatase and osteocalcin).

Concerning bone levels, 30 days following the ovariectomy, bone mineral content has been found to decrease. Through histomorphometry, there is a decrease in the trabecular bone volume; however,

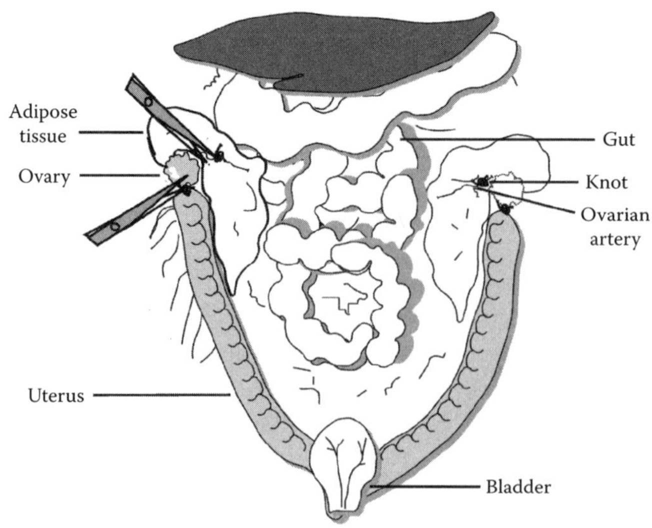

FIGURE 29.2 Shown are the location of the ligatures and area of cutting before ablation (right ovary, in gray) and after ablation (left ovary, in white).

TABLE 29.1
Weight of Uteri of Rats with Ovariectomy (OVX) and Controls (Sham) 30 Days after Surgery

Group	Uterus Weight (mg/100 g Body Weight)
Sham	140.0 ± 20.0
OVX	32.3 ± 1.1*

Student's T test independent data, * $p < 0.001$.
Note: Results are expressed as mean ± SEM.

there is an increase of the surface covered by osteoclasts and osteoblasts, osteoclasts number, speed of the osteoid surface formation, speed of mineral apposition, osteoid thickness, and the delay time in mineralization. It is interesting to note that the younger female rat is more responsive than the aged rat (6 to 12 months or more). The proximal tibia and the lumbar vertebra are the most used sites for the study of osteoporosis. Both areas respond in a similar way to the deficiency of estrogens, although the loss of bone is different. In the tibia, the bone loss is the result of the decrease in the number of trabeculas; in the vertebra, it is the consequence of the decrease in the trabecular thickness.[3]

Precaution: It can happen that the ablation of the glands is not complete. In this case, the rat will not present a failure in the ovarian function. If the ovariectomy was successful, ovarian tissue should not be found and atrophy of the uterus should be observed. To assess the atrophy, the uterus can be excised and weighed.

OVX produces hyperphagia in animals. Therefore, it is a factor that must be controlled in chronic experiments. It is recommended to work on a group in which a simulated surgery was performed (sham-operated group). Animals from this group must be fed with *ad libitium* food and the intake of the food must be controlled daily. Using this data, the amount of food of the OVX group must be adjusted. Even with this precision, OVX animals can present overweight. Calcium, phosphate, and vitamin D contents of the diet must be controlled. Recommended values are 1 wt% calcium, 0.3 wt% phosphate, and 3.0 U vitamin D/g of diet. High calcium content decreases the effects of ovariectomy in bones.[2]

In this case, hopefully the functions of estrogen in OVX rats can be reestablished. The hormone must be administered intradermically, in a 15 μg/kg/day dose. Through this administration, just some of the bone parameters already mentioned are re-established.

Experimental Results

Eight 70-day-old female Sprague Dawley rats were subjected to ovariectomy (OVX group) and 8 rats were subjected to a simulated surgery (sham group). The weights of the uteri 30 days after the surgery are shown in Table 29.1.

REFERENCES

1. Wronsky, I.J., P.L. Lowry, C.C. Walsch, and L.A. Ignaszewsli. 1985. Skeletal alterations in ovariectomized rats. *Calcif. Tiss. Int.* 37: 324–328.
2. Kalu, D.N. 1991. The ovariectomized rat model of postmonopausal bone loss. *Bone Miner.* 15: 175–192.
3. Thompson, D.D., H.A. Simmons, C.M. Pirie, and H.A. Ke. 1995. FDA Guidelines and animal models for osteoporosis. *Bone* 17: 125S–133S.
4. Bonjour, J.P., P. Ammann, and R. Rizzoli. 1999. Importance of preclinical studies in the development of drugs for treatment of osteoporosis : A review to the WHO guidelines. *Osteoporos Int.* 9(5): 379–393.
5. Miller, S.C., B.H. Bowman, and W.S. Jee. 1995. Available animal models of osteopenia: Small and large. *Bone* 17(4): 117S–123S.

30 Thyroparathyroidectomy

Verónica E. Di Loreto and Alfredo Rigalli

INTRODUCTION

The ablation of thyroid and parathyroid glands is performed through the thyroparathyroidectomy (TPTX). It is a low complexity surgery that can be performed in 20 minutes and it presents high chances of success. The thyroid gland lies on the lateral surface of the trachea just below the larynx and covers four or five tracheal rings. The two thyroid lobes are joined by an isthmus. The gland is covered by three muscles: sterno-thyroid, thyro-hyoid, and sterno-hyoid. The recurrent laryngeal nerve is situated between the trachea and the lobes of the gland.

UTILITY

This surgery is useful in studying the role of the thyroid or parathyroid hormone in different physiological or physiopathological processes. It is also used to evaluate if a drug acts directly on a process or through the modification of the thyroid or parathyroid function. For example, if a drug produces hyperphosphaturia, this may be caused by its direct action on the kidneys or by the release of parathyroid hormone (PTH), which acts on the kidneys. The administration of the drug to TPTX and intact rats will provide evidence of its mechanism of action.

Materials

General anesthetic
Thermostatized rat board
Ring-handled preparation scissors
Cotton pliers (2)
Adson forceps
Artery forceps (2)
Needle holder
Antibiotics
Analgesics
Saline solution
Iodopovidone
Injection needle 15 × 0.5
Scalpel (no. 15)
Suture needle curved ½ – 5 to 10 mm
Suture thread
Cotton buds

PROCEDURE

The surgery must be performed using general anesthetic. Once the rat is placed in a supine position, a median incision is made on the skin at the level of the trachea cutting toward the edges of the jawbone (Figure 30.1).

Then, the muscle that covers the middle line is divulsed and separated with the artery forceps to expose the thyroid gland (Figure 30.2).

Once located, each lobe of the gland is held tightly with a forceps (preferably with an Adson forceps) and pulled gently upward and laterally, releasing it with a scalpel or with the bevel of a 15 × 0.5 injection needle. A nerve can be see on both sides of the gland. Care must be taken to prevent damage to the gland, which can cause a glottis spasm that can bring about the death of the animal through asphyxia.

After the ablation of the gland, the edges of the muscles are joined together and sutured. Finally, the skin is sutured.

Special Considerations

If the experiment only involves studying the effects of the parathyroid hormone, thyroxine should be administered to the animal through drinking water (0.5 μg/100 g body weight per day).[1]

The success of the thyroid and parathyroid ablation is verified through the measurement of the calcemia after 48 to 72 hours from the extraction. For example, IIM/Fm strain subline "m" rats reach a 7 to 7.5 mg/dl calcemia without the presence of tetany.[1]

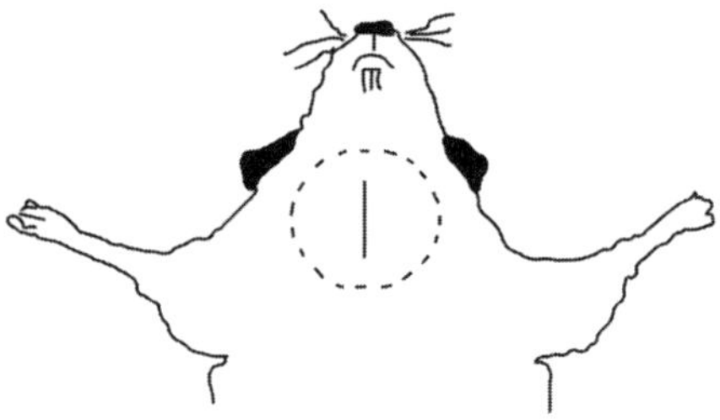

FIGURE 30.1 Cutting area for the ablation of the thyroid gland is illustrated.

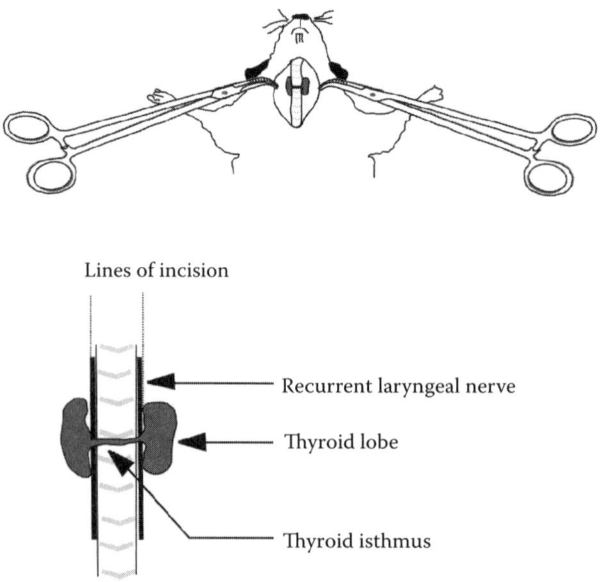

FIGURE 30.2 Shown is the location of the thyroid gland and recurrent laryngeal nerve related to the trachea.

TABLE 30.1
Values of Calcemia (mM), Phosphatemia (mM), Phosphaturia (mg/day 100 g Body Weight), and Parathyroid Hormone (ng/dl) in Control and Diabetic Rats

	Control		Diabetic	
	Intact	**TPTX**	**Intact**	**TPTX**
PTH	15.6 ± 0.3	Not detected	52.7 ± 1.8*	Not detected
Calcemia	2.43 ± 0.4	1.91 ± 0.06*	2.41 ± 0.2	1.83 ± 0.09*
Phosphatemia	1.92 ± 0.07	2.43 ± 0.15*	1.88 ± 0.39	1.73 ± 0.46
Phosphaturia	13.7 ± 1.0	10.5 ± 6.0*	28.7 ± 3.6*	41.5 ± 5.5

* Shows significant differences with respect to intact control rats $p < 0.05$.
Note: The values are expressed as mean ± SD.

Experimental Results

The ablation of the thyroid and parathyroid glands was performed in a control group (n = 6) and in diabetic rats (n = 6). Moreover, another control group (n = 6) and diabetic rats (n = 6) had a simulated surgery (sham-operated rats) and were used as intact animals. Rats were placed in individual metabolic cages for 10 days to perform efficiency of food conversion measurements. With the purpose of maintaining the thyroid function, the drinking water of the TPTX rats was provided with L-thyroxine, with a concentration that provides 0.5 μg/100 g body weight per day.

Urine was collected to determine phosphaturia and, at the end of the experiment, the anesthetized animals were sacrificed by heart puncture. Blood was withdrawn for the measurement of PTH, calcium, and phosphate. The lack of PTH produces a decrease of phosphaturia in normal rats, but this effect was not observed on diabetic rats due to competition between phosphate and glucose for the mechanism of transport. The results are shown in Table 30.1.

REFERENCE

1. Locatto, M.E., V. Di Loreto, M.C. Fernández, D. Caferra, and R.C. Puche. 1997. Relative weight of glucosa, insulin and parathyroid hormones in the urinary loss of phosphate by chronically diabetic rats. *Acta Diabetol.* 34: 211–216.

31 Parathyroid Glands

Maela Lupo and Alfredo Rigalli

INTRODUCTION

Parathyroid glands produce the parathyroid hormone. This hormone is involved in calcium homeostasis and has direct effects on bone and kidneys. It also has indirect effects on the intestine through calcitriol. In conjunction with other hormones and cytokines, it controls bone remodeling, renal handling of phosphate, and renal calcium reabsorption.

Parathyroidectomy (PX) is the ablation of the parathyroid glands, which are located in the thyroid gland and can be distinguished because they are two white spots inside the thyroid. The thyroid gland lies on the lateral surface of the trachea just below the larynx and covers four or five tracheal rings. PX is a low complexity surgery and can be performed in approximately 30 minutes. A more detailed explanation of the location of the thyroid gland can be found in Chapter 30.

Utility

The main advantage of this surgery, in contrast to the thyroparathyroidectomy, is that with PX, one can study the effects of parathyroid hormone with intact thyroid function. Furthermore, it is recommended for the induction of hypoparathyroidism, a pathology caused by inadequate parathyroid hormone (PTH) secretion. Other uses of PX include modification of bone remodeling and renal handling of phosphate.

Materials

General anesthetic
Thermostatized rat board
Ring-handled preparation scissors
Ring-handled iris scissors
Cotton pliers (2)
Adson forceps
Rat-tooth forceps
Artery forceps (2)
Needle holder
Microforceps (2)
Stereo microscope
Antibiotics
Analgesics
Saline solution
Iodopovidone
Scalpel (no. 15) or injection needle 25 × 0.8
Suture needle curved ½ – 5 to 10 mm
Suture thread
Cotton buds

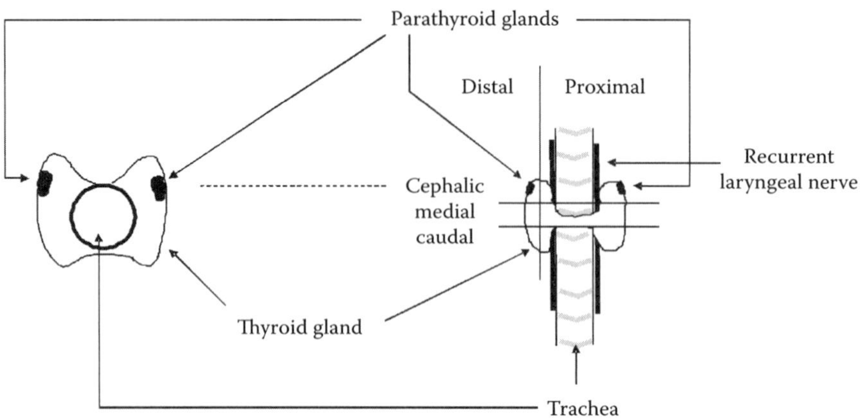

FIGURE 31.1 Location of thyroid and parathyroid glands is illustrated.

Procedure

The same procedure described in the thyroparathyroidectomy (Chapter 30) is followed in this surgery to find the thyroid gland. Once the thyroid gland is located, one searches for the parathyroid glands. In order to readily locate the parathyroid glands, the thyroid gland can be divided into three sections: cephalic, medial, and caudal. Having done this, parathyroid glands are found situated in the cephalic section distal to the trachea (Figure 31.1). They are two white spots of 0.5 to 1 mm in diameter. When the parathyroid glands are identified, microforceps and a beveled 25 × 0.8 injection needle or a scalpel should be used to remove the glands.

Special Considerations

The success of the parathyroid ablation is verified through the measurement of the calcemia after 48 to 72 hours from the extraction. In general, plasma calcium levels lower than 7.5 mg/dl indicate a successful surgery. To avoid tetany due to severe hypocalcemia, 10 g/l calcium gluconate in the drinking water should be administered. It is advisable to put two drinking water containers, one with tap water and another with 10 g/l calcium gluconate.

Experimental Results

The ablation of the parathyroid glands was performed to produce rats with low bone remodeling. This was completed in two groups: rats with PX surgery (PX, n = 8) and rats with simulated surgery (sham, n = 8). Blood was obtained before and 72 hours after the surgery for the measurement of calcemia (Table 31.1)

TABLE 31.1
Calcemia (mg/dl) in Parathyroidectomized and Control Rats

	Before Surgery	72 Hours After Surgery
PX	9.7 ± 0.9	3.8 ± 1.9
Sham	9.8 ± 0.7	10.0 ± 0.5

Note: The values are expressed as mean ± SD.

Section VIII

Kidneys

32 Measurement Techniques of Renal Parameters

Verónica E. Di Loreto and Alfredo Rigalli

INTRODUCTION

The first step in urine formation is plasma filtration. Generally, between 20 to 30% of the renal plasma flow (RPF) goes through the glomerular capillary to the Bowman's capsule to produce an ultrafiltrate free of proteins. The composition and volume of this ultrafiltrate are then modified by processes of reabsorption and secretion of substances in the renal tubules. Therefore, being able to study these processes is extremely important in understanding the physiology and physiopathology of the kidney.

RENAL CLEARANCE OF SUBSTANCES

The concept of the clearance of substances in the kidney is central in renal physiology as it provides a way to evaluate the excretion of substances through the kidney.

Renal clearance is the volume of plasma that is equal to the amount of substance being eliminated through urine per minute. Clearance can take values from zero (in substances that are not eliminated through urine) up to a value equal to renal plasma flow (in those substances completely eliminated from the blood through the kidney).

Given an X substance, the rate of urinary excretion is defined as $U_x.Vu'$ (expressed in the amount of the substance/min) where U_x is the urinary concentration of X, and Vu' is the urinary flow. If P_x is the plasma concentration, the ratio $U_x .Vu'/P_x$ represents the volume of plasma that contains the same amount of X substance that is eliminated in the urine per minute. It is called "renal clearance of X" and is a virtual volume, not a real one. It is an empiric measure of the kidney's ability to remove a substance from the blood and deliver it into the urine.

In a general way, for a given substance X, the clearance concept can be formulated as:

$$Cl_X = \frac{V_u' \times \acute{U}_X}{P_X} \tag{32.1}$$

The greatest advantage of clearance measurement is that it only requires the collection of urine and plasma. Among the different kinds of renal clearances, there are two that are very prominent: (1) clearance of inulin and (2) clearance of p-aminohippuric acid, which allows the measurement of the glomerular filtration rate and renal plasma flow, respectively.[1,2]

GLOMERULAR FILTRATION RATE (GFR)

The glomerular filtration rate is a good index of renal function and a significant concept to understand renal physiopathology. Its value can be estimated through clearance methods. The ideal substance must be biologically inert and nontoxic, quantifiable in plasma and urine with a high degree of precision, freely filtrated through the glomerulus, and not reabsorbed or secreted by the tubules. In this case, the clearance of this substance is an accurate measurement of its glomerular filtration, i.e., it measures the volume of plasma filtered through the glomerulus in 1 minute. The most studied and widely known substance that combines these characteristics is inulin (fructose polymer), and its clearance is an accurate measure of GFR and can be used to estimate this rate. Therefore, for this substance and, according to mass balance between plasma and urine, it can be said:

Filtered quantity = Quantity excreted through urine

$$GFR \times P_{In} = V_u' \times U_{In} \tag{32.2}$$

where GFR = glomerular filtration rate, P_{In} = inulin plasma concentration, V´u = urinary flow, and U_{In} = urinary concentration of inulin. Then,

$$GFR = \frac{V_u' \times U_{In}}{P_{In}} \tag{32.3}$$

that is the inulin clearance.[1,2]

Renal Plasma Flow (RPF)

For some substances, tubular secretion can be very relevant. An example of this is p-aminohippuric acid (PAH), which is an organic acid that is not produced in the body. This substance is nontoxic and not metabolized by extrarenal tissues, it is neither synthesized nor destroyed by the kidney, and it is only restricted to plasma, which validates the mass plasma–urine balance. The PAH in low plasma concentrations is so efficiently secreted by the proximal tubule that the blood that leaves the kidney is free of PAH. Therefore, its renal clearance provides an approximate measure of RPF. Then, according to mass balance, it can be said:

Quantity entering to kidney through the blood = quantity excreted from kidney through the urine.

$$RPF \times P_{PAH} = V_u' \times U_{PAH} \tag{32.4}$$

Then,

$$RPF = \frac{V_u' \times U_{PAH}}{P_{PAH}} \tag{32.5}$$

that is the clearance of PAH (Cl_{PAH}).[1,2]

MEASUREMENT OF GLOMERULAR FILTRATION RATE AND RENAL PLASMA FLOW

Materials

General anesthetic
Thermostatized rat board

Ring-handled preparation scissors (2)
Cotton pliers (2)
Adson forceps
Needle holder
Spatula
Peristaltic pump for perfusion
Saline solution
Iodopovidone
Heparin
Drugs: inulin, PAH, mannitol
Catheters PC-40 for artery and PC-50 for vein
Catheter PC-35 or similar for bladder
1-ml syringe
Injection needle 25 × 0.8
Suture needle ½ – 5 to 10 mm
Suture thread
Cotton buds

Procedure

The measurement of these two renal hemodynamic parameters is performed in the following way. Once the animal is anesthetized, with the aim of obtaining blood samples and perfusing the animal, heparinized catheters are placed in the femoral artery and the vein, respectively. Also, a catheter is placed into the bladder to collect urine samples. Once these procedures are performed, the animal is injected with a single dose of 18 mg inulin and 6 mg PAH (priming dose) dissolved in saline solution (2 ml) and at a pH = 7.4 through the vein, to rapidly increase the plasma concentrations. Subsequently, the animal is perfused with an isotonic solution composed of 9 g/l inulin, 3 g/l PAH, and 50 g/l mannitol at a pH = 7.4. Perfusion is performed at a rate of 0.1 ml/min. After 45 minutes, when plasma levels of inulin and PAH reach the steady state, urine is collected during different periods of time determined for the experimental design, its volume is measured (to estimate Vu´), and artery blood samples are extracted in the middle of each period. Inulin and PAH are determined in plasma and urine samples. Then, inulin clearance and PAH clearance are calculated with the equations stated above (Equation 32.3 and Equation 32.5). Both of them are expressed in ml/min.

The inulin and PAH clearances can be individually determined if the experimental design requires it.

Experimental Results

With the purpose of measuring the GFR, experiments of perfusion are performed in control and alloxan diabetic animals.[3] Sixteen control animals and 30 diabetic animals were anesthetized with urethane intraperitoneally and their femoral arteries and veins were catheterized. Also, a bladder cannulation was performed for urine collection.

The animals were perfused through the catheter placed in the vein with the described solution. Perfusion flow was fixed at 0.1 ml/min. After 45 minutes, urine was collected twice in 1 hour. The blood samples were extracted through a catheter placed in the artery in the middle of each period. Two hundred μl of blood were extracted per period. In blood as well as in urine, the levels of inulin and glucose were determined. The results are displayed in Table 32.1.

RENAL BLOOD FLOW MEASUREMENT

If the hematocrit (Hct) is measured, RPF can be converted to renal blood flow (RBF) through the following equation:

$$RBF = \frac{RPF}{1 - Hct} \tag{32.6}$$

The described conventional clearance technique provides a global measure for GFR and RPF. Although PAH clearance is the most commonly used method in the practice to measure RBF, it provides little information about the distribution of blood between the cortex and the medulla. Apart from the clearance techniques, there are other methods to measure total RBF, which are described in Chapter 35.

MEASUREMENT OF THE FILTRATED LOAD OF A SUBSTANCE

When the reabsorption or secretion of a substance is studied, it is necessary to know the rate at which it is filtered and excreted. For those substances that are freely filtered through the glomerulus and whose concentration in the ultrafiltrate is, therefore, equal to the plasma concentration, the filtered load per unit of time (FL_X) can be then calculated as the product of the plasma concentration (P_X) and the glomerular filtration rate (GFR):

$$FL_X = P_X \times GFR \tag{32.7}$$

MEASUREMENT OF THE RENAL REABSORPTION OF A SUBSTANCE

The excreted amount of a substance X that suffers tubular reabsorption, that is to say, the urinary concentration of X per volume minute of urine ($U_x \times V'_u$) is always smaller than the filtered load. The difference between them determines the amount of reabsorbed X. The net reabsorption rate of X (T_X) is defined as the difference between the filtered load and the excretion rate, and can only be measured indirectly:

$$T_X = P_X \times GFR - U_X \times V'_u \tag{32.8}$$

TABLE 32.1
Glycemia and Clearance of Inulin in Normal and Diabetic Rats (a, b, and c Show Significant Differences as Regards Control Rats p < 0.05, p < 0.01 and p < 0.001, Respectively)

	Glycemia (mg/dl)	Inulin Clearance (ml/min/100g Body Weight)
Control rats (n = 16)	105 ± 6	0.97 ± 0.04
Diabetic rats hypofiltering (n = 30)	442 ± 33 [b]	0.55 ± 0.07 [c]
Diabetic rats normofiltering (n = 4)	339 ± 23 [a]	0.87 ± 0.04

Note: The data are expressed as mean ± SD.

Experimentally, it is demonstrated that, when the filtered load of a substance is extremely low, its excretion is undetectable because it is totally reabsorbed. Increasing the filtered load, the reabsorption rate is raised to a maximum, denominated by "tubular maximum of X" or Tm_x (this only occurs for substances that are actively reabsorbed). Any subsequent increase in the filtered load as a result of an increase in the plasma concentration is completely eliminated.

The fractional tubular reabsorption of X (TRx) is defined as the filtered fraction of X that is reabsorbed and the fractional excretion of X (FEx) as the filtered fraction of X that is eliminated.

A practical way to perform an estimate of the tubular reabsorption of a substance is based on the ratio substance clearance/inulin clearance (Cl_x/Cl_{In}).

The fractional reabsorption of X is easily obtained using measures of the inulin concentration and the studied substance in plasma and urine collected simultaneously. The TRx equals:

$$TR_X = \frac{\left(FL_X - U_X \times V_u'\right)}{FL_X} = 1 - \frac{U_X \times V_u'}{FL_X} = 1 - \frac{Cl_X}{Cl_{In}} \qquad (32.9)$$

The Cl_x/Cl_{In} ratio represents the fractional excretion of X.[4]

Materials and Procedure

They are similar to GFR measurement described above. The only difference is the addition of the substance of interest to the solution for performing the perfusion. See experimental results for an example.

Experimental Results

Tubular Reabsorption of Phosphate

The experiments were carried out with male, 7-week-old IIM/Fm rats, weighing 167 ± 19 g (mean ± SD). Under general anesthesia, the catheters were placed into the right femoral artery (to obtain blood samples), into the left femoral vein (to perfuse the animal), and into the bladder (to collect urine samples). The animals were perfused at the rate of 0.1 ml/min with an isotonic inulin solution (inulin 9 g/l, mannitol 50 g/l, and phosphate 1.1 mM, pH = 7.4). An initial perfusion period (approximately 30 minutes) was necessary until the plasma phosphate level became stable (2.7 to 3.0 mM).

Inulin and phosphate were measured in plasma and urine. The data were used to calculate the glomerular filtration rate (GFR: inulin clearance, ml/min), the fractional excretion (FE_{PO4}), and the tubular reabsorption of phosphate TR_{PO4} (Table 32.2).

In conclusion, almost all the phosphate that is filtered by the glomerulus is reabsorbed in tubules (approximately 95%).

MEASUREMENT OF EXTRACELLULAR VOLUME (ECV)

The volumes of several fluid compartments of the body, among them the extracellular volume, can be measured using the dilution principle. In theory, using this method, we can measure the volume of any compartment if there is a suitable marker substance. This substance, after injection into the compartment, should distribute homogeneously without being excreted in urine. In practice, a known amount of a marker substance (test substance) is injected into that compartment, it is allowed to penetrate uniformly, and its concentration is measured from a sample drawn from the compartment. The volume in which the substance was distributed is calculated with the following equation:

TABLE 32.2
Tubular Reabsorption of Phosphate

Parameter	Value
GFR (Cl_{IN}, ml/min)	1.96 ± 0.24
V'_u (ml/min)	0.07 ± 0.03
P_{PO4} (mg/dl)	7.43 ± 0.38
U_{PO4} (mg/dl)	10.74 ± 3.62
FL_{PO4} (mg/min)	0.146 ± 0.09
T_{PO4} (mg/min)	0.138 ± 0.07
FE_{PO4}	0.052 ± 0.01
TR_{PO4}	0.948 ± 0.03

Note: The results are expressed as the mean ± SD.

Volume = Amount of (known) injected substance/substance concentration

The ECV is particularly difficult to measure because there is not an ideal substance that only penetrates the extracellular compartment without penetrating the intracellular compartment. Besides, the substance should be excreted slowly in comparison with the rate at which it distributes through the extracellular space and must fulfill the following conditions:

- To be nontoxic
- To distribute uniformly within the compartment
- Not to be metabolized during the period the test is performed
- Not to alter the volume of the compartment being measured

In practice, it has been impossible to find a substance satisfying all the above-mentioned requirements; therefore, it is not a precise measurement. As a consequence, ECV is not calculated, but estimated.

Generally, two kinds of test substances are used: (1) saccharids, such as insulin, mannitol, and sucrose; and (2) ions, such as thiocyanate, thiosulfate, sulfate, bromide, and chloride. These substances have different rates of penetration into the various components of extracellular space, so their equilibrium phases are not equal. As a consequence, the ECV value can differ somewhat when using one or another.[1]

Techniques

When the dilution principle is applied, the test substance injected in plasma is being excreted through the kidneys. As a consequence, the plasma concentration of the test substance decreases continuously, making it difficult for the application of the dilution principle in its standard forms. To overcome this difficulty, two procedures were developed, which depend on the rate of the renal excretion of the test substance: (1) method of the single dose injection, for substances whose renal excretion rate is slow; and (2) method of continuous infusion, for substances whose renal excretion rate is rapid.

Though inulin is quickly excreted by the kidney, in our experience, it can be used to perform both methods and gives comparable results.

Single Dose Injection Method

When a known amount of the test substance is injected in plasma, two factors determine its plasma concentration: renal excretion and the possible penetration in other compartments. When the injected substance reaches its uniform distribution, the rate of decrease of plasma concentration will be constant and will equal its rate of renal elimination.

Materials

General anesthetic
Thermostatized rat board
Ring-handled preparation scissors (2)
Cotton pliers (2)
Adson forceps
Spatula
Catheters PC-40 for artery and PC-50 for vein
1-ml syringe
Injection needle 25 × 0.8
Saline solution
Heparin
Inulin
Suture thread
Cotton buds

Procedure

With the rat under general anesthesia, catheterization of the femoral artery and vein must be performed. After this, a single dose of inulin is administered in a known amount (18 mg dissolved in 2 ml of saline solution, pH = 7.4) through the femoral vein and a sample of blood is immediately drawn through the femoral artery. Then, several consecutive extractions are performed at 3, 5, 10, 15, 20, 30, and 40 minutes. Inulin concentration in plasma is measured (preferably by duplicate). To find the volume of the compartment where the inulin has penetrated, it is necessary to calculate the plasma concentration that should have been obtained if it penetrates uniformly and instantaneously through the ECV and has not yet been eliminated through the kidney. This is called instantaneous concentration and it is graphically determined as follows:

Using the measured values of plasma inulin concentration, the logarithm of plasma concentration against time is plotted (Figure 32.1) and the instantaneous concentration is obtained by extrapolating the linear portion of the curve (which implies constant renal excretion) back to zero. To estimate the ECV, the initially injected amount of inulin is divided by the instantaneous concentration of inulin. The procedure can be used with different test substances.

Experimental Results

The extracellular volume was estimated in a group of 6 rats (strain IIM/Fm subline "m") using the principle of dilution, and the single dose injection method with inulin as the test substance.

Using the protocol described above, the instantaneous concentration was graphically determined and the ECV was estimated. It was expressed per 100 g body weight. The value of the extracellular liquid obtained was 22.18 ± 1.60 ml/100 g body weight.

Continuous Infusion Method

Material

The materials are similar to those in a single dose injection method (see above), with the addition of a peristaltic pump.

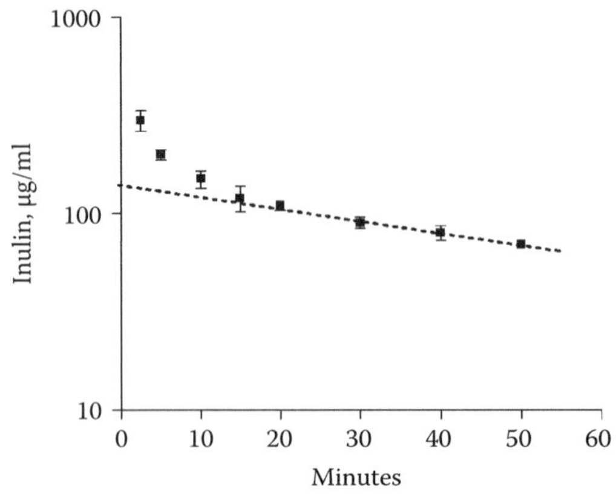

FIGURE 32.1 Determination of inulin instantaneous concentration is illustrated.

Procedure

With the rat under general anesthesia, catheterization of the femoral artery and vein, and a vesical catheterization are performed. Through the vein, a bolus of inulin (priming dose, 18 mg in 2 ml of saline solution, pH = 7.4) is injected to raise its plasma concentration. Then, an inulin solution (9 g/l in saline solution, pH = 7.4) is administered with a perfusion pump at a rate similar to its renal excretion (around 0.1 ml/min) during approximately 2 hours in order to stabilize the plasma concentration. To corroborate this constancy, it is necessary to draw some plasma samples during this equilibration period (e.g., at 1.30 and 2 hours) and to determinate inulin concentration in them. Then, the bladder is emptied and the administration is stopped. Urine is collected at this time until all the inulin contained in the body is eliminated (approximately 6 to 7 hours). Measuring inulin concentration in urine at different periods of time can corroborate this. The total amount of inulin contained in the volume of urine collected represents the quantity of the inulin that was retained in the body and yielded the plasma concentration when the infusion was stopped. This value is divided by the constant plasma concentration of inulin and the ratio gives, as a result, the ECV into which the inulin was evenly distributed.

Experimental Results

A group of 6 rats (strain IIM/Fm subline "m") was anesthetized and the continuous infusion method was performed as described above. The ECV value obtained was 24.29 ± 5.13 ml/100 g body weight. As one can see, the results were no different from those obtained through the unique inulin injection method (Student's t test, $p > 0.05$).

REFERENCES

1. Koushanpur, E., and W. Kriz. 1986. Renal clearance: Measurements of glomerular filtration rate and renal blood flow. In *Renal physiology: Principles, structure and function*, 2nd ed. Heidelberg: Springer-Verlag, pp. 96–111.
2. Greger, R. 1996. Introduction to renal function, renal blood flow and the formation of filtrate. In *Comprehensive human physiology*, vol. 2, R. Greger and U. Windhorst (eds.). Berlin/Heidelberg: Springer-Verlag, pp. 1469–1487.
3. Di Loreto, V., H. Moreno, R. Puche, and M. Locatto. 2004. Severe hyperglycemia: A determinant for hypofiltration in alloxan diabetic rats. *Acta Diabetol.* 41: 56–62.
4. Bijvoet, O.L.M. 1976. The importance of the kidney in phosphate homeostasis. In *Phosphate metabolism kidney and bone.*, L. Avioli, Ph. Bordier, H. Fleish, S. Massry, and E. Slatopolsky (eds.). Paris: Armour Montagu, pp. 421–474.

33 Kidney Isolation and Perfusion

Verónica E. Di Loreto and Alfredo Rigalli

KIDNEY ISOLATION AND PERFUSION

Utility

The isolated kidney is appropriate for the study of several physiological and biochemical aspects of the renal function. The main advantage is that it is possible to work under conditions in which the variables can be modified in a controlled way, eliminating the systemic influences, while the anatomic, biochemical, and functional characteristics are preserved.[1] Therefore, this model allows the study of the direct effects of different substances on the kidney in the absence of extrarenal factors that interfere in the information obtained in experiments carried out *in vivo*.

The glomerular functions and all the transport functions that take place exclusively in the proximal tubule are preserved in the isolated kidney. On the other side, many transport functions that depend partially or completely on the integrity of the functions of the distal nephron are severely distorted. Also, the capacity of concentration and dilution of urine are damaged.

Experiments performed with micropuncture in the isolated kidney confirm the functional characteristics of the experimental surgical model.[1]

As regards the hemodynamic characteristics, the filtration fraction is almost negligible. This abnormality is not due to an intrinsic defect of the isolated kidney, but to the hydrodynamic properties of the perfusate, that is to say, to its high renal perfusion flow. Therefore, isolated kidney perfusion is not an adequate model for the renal hemodynamic study.

The technique is useful in the study of issues that cannot be resolved through experimentation with the intact animal, especially when phenomena produced almost exclusively in the proximal tubule are analyzed, such as tubular reabsorption of organic compounds and phosphate. The advantage of this technique is that some experimental maneuvers that are difficult or impossible to perform in *in vivo* experiments are easily achieved with the isolated kidney.

The fact that the experimental variable can be controlled while the functional integrity of the organ is preserved is an advantage with respect to other *in vitro* experiments, such as kidney slices.[1]

To sum up, isolated kidney perfusion is an experimentally useful model through which some aspects of the function and metabolism of the kidney can be studied. When the experimental design is appropriate and the results are sensibly interpreted, it can turn into an important experimental tool in the study of renal physiology and physiopathology.

Materials

General anesthetic
Thermostatized rat board
Ring-handled preparation scissors
Iridectomy scissors
Cotton pliers (2)
Adson forceps
Artery forceps (6)
Needle holder

Spatula
Peristaltic pump
Thermostatic bath
Carbogen tube
Filters
Tygon hose
Pressure translator or mercury manometer
Flow meter
Stereomicroscope
Catheters: (1) for ureter: PC-40 catheter approximately 6 cm long or PC-50 catheter of the same length, with a PC-10 of 2 cm connected to the end that is used to catheterize; (2) for aorta artery: PC-100 or K35 catheter with approximately 3 cm of a PC-75 catheter inserted on its end; and (3) for vein cava: K35 catheter
Injection needle 25 × 0.8
Saline solution
Heparin
Iodopovidone
Suture needle ½ – 5 to 10 mm (2)
Suture thread
Cotton buds

Procedure

After the animal is anesthetized, the femoral artery is catheterized to administer heparinized saline solution (150 IU: 0.03 ml heparin 5000 IU/ml in 0.5 ml of saline solution). Shurek and Alt[2] developed the technique used. Subsequently, a median laparotomy is performed from pelvis to approximately the sternum, wide enough (5 to 7 cm) to work comfortably. Once the cut is performed, the artery forceps are used to keep the field wide open. The intestinal loops are moved to the left and the right kidney is immediately detected. Also visible is the mesenteric artery, the celiac artery, and the right renal artery (Figure 33.1).

It is important that the surgical area is well cleared to work comfortably and avoid excessive manipulation of the area. The isolation of the right kidney is preferable because of its easier accessibility.

As a first step, the right ureter is identified and bound proximally to the bladder. Although its cannulation for urine collection is preferably performed at the end of the surgery, the binding favors its subsequent location and cannulation, as the urine accumulated in it will enlarge its diameter. Even though some execute this cannulation at the beginning of the technique, we prefer to perform it at the end to avoid the binding from releasing because of the manipulations of the surgery. The cannulation is performed through the insertion of a PC-40 or PC-50 catheter connected to a PC-10 catheter of approximately 2 cm length.

The next step consists of identifying all the necessary vessels to be clamped: left renal artery and vein, right adrenal artery, upper mesenteric artery, and celiac artery (see Figure 33.1). Then, the areas are cleared from connective tissue and threads are placed around each vessel to clamp them. Also, the abdominal aorta and the lower vein cava must be identified. It is important to perform this identification in two areas: from above and from under the anastomosis between the upper mesenteric artery and the right renal artery. Above the anastomosis, both vessels are bound. These binds, together with the ones performed in the beginning, are to allow the solution of perfusion to go only to the kidney and not to extrarenal areas. In the case of the vein cava, the aim of its upper binding is to force the renal flow to follow the cannulation direction.

Subsequently, the aorta is clamped, distal to the renal artery, and the circulation is momentarily stopped. The clamp can be performed with a thread passed around the vessel, pulling softly in the opposite direction of the blood flow. With the iridectomy scissors, an incision is performed in the

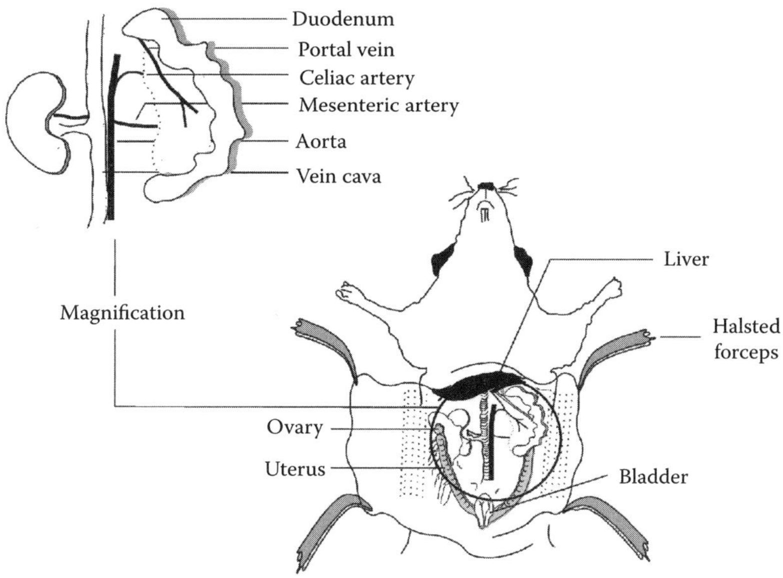

FIGURE 33.1 Surgery field for isolated kidney is illustrated.

aorta, under the clamp, and a catheter is inserted, pushing it softly to the origin of the right renal artery. Once the catheter is inserted, releasing the clamp reinstates the aortic circulation, and the perfusion of the organ starts immediately. (It is possible to check that the isolation is correctly performed because the kidney loses color immediately.)

Then, the lower vein cava is catheterized under the right renal vein, the organ can be cleaned, and the drain of the reservoir starts through perfusion in order to establish a closed circuit.

There is an alternative for the catheterization of the right renal artery described by other authors.[3,4] This consists of catheterizing the renal artery via the upper mesenteric artery. A small incision is performed in that artery and a 25 × 1 blunt injection needle is placed through the mesenteric artery, passing through the aorta to the renal artery. Then, the catheter is bound to the renal artery as well as to the mesenteric artery. Finally, the catheterization of the ureter is performed. Because of the small diameter of the ureter, a stereomicroscope must be used in this procedure.

Precautions

As deduced from the technical description, it is a complex surgical procedure. Excessive vessel manipulation can cause an important vasoconstriction, anoxia, and an experimental model that will provide nonreliable results.[4] Also, excessive bleeding can make catheterization difficult.

Another factor to consider is maintaining the organ temperature between 35 and 37°C. For this, a thermostatic board or an infrared light lamp can be used. The latter can lead to the dehydration of the organ, thus renal humidity must be maintained by covering the organ with a piece of cotton soaked with saline solution.

Once the catheterization of the ureter, renal, or aorta artery and vein cava are performed, and the perfusion started, some authors decapsule and remove the kidney, placing it in a thermostatic chamber at 34 to 37°C.[3,4]

Solution of Perfusion

Its composition is extremely important for the stable functioning of the isolated kidney. The isolated kidney is perfused with a solution that contains electrolytes (e.g., Krebs–Ringer bicarbonate (KRB) buffer), a mixture of amino acids to improve the stability of the experimental model,[5] and glucose as an energy source, to which an oncotic agent, such as albumin or dextran, is added. The oncotic agent generates a "physiological" coloidosmotic pressure, which is one of the key factors in controlling the glomerular filtration rate (GFR), tubular reabsorption of water and sodium, and the content of water in the interstitial renal tissue.[3]

The composition of the solution of perfusion used in our laboratory includes: KRB buffer pH = 7.4, 300 mosmole/l; content: dextran at 2% w/v (PM 70,000), glucose 5 mM, inulin 0.6 mg/ml, sodium piruvate 5 mM, lactic acid 5 mM, and the following amino acids: L-glutamate 0.5 mM, L-glycine 2.3 mM, L-cysteine 0.5 mM, and L-arginine 0.6 mM. This solution must be bubbled with a mixture of 95% O_2 and 5% CO_2.

The volume of the solution can vary depending on the experimental needs. In our laboratory, an initial volume of 100 ml is used. This volume is maintained constant to recover the volumes extracted from the solution of perfusion and the urinary volume losses.

The composition of the solution may vary according to each research group. Some differences are:

1. Different amino acids combinations
2. Coloidosmotic agent: use of bovine serum albumin (BSA) instead of dextran.

The albumin concentration commonly used is 60 to 65 g/l. The molecular weight of dextran must be ranged between the 60,000 and 90,000 Da. BSA is a much used oncotic agent. The functional parameters obtained are stable, but the GFR values are lower than those found *in vivo*.[1,4]

Dextran does not bind other compounds, as albumin does. This characteristic makes dextran more useful for pharmacokinetic investigations with drugs.

In the informative work of Wang et al.,[3] they describe the use of dextran to replace a great amount of BSA and establish a more reliable and economical model. In this work, the performance of the kidney perfused with a mixture of BSA/dextran is compared with the use of BSA as an oncotic agent, and they obtained a higher renal viability with the mixture. The values of GFR obtained are similar to those found in the animal *in vivo*. In this way, the result is a model that is recommended for its stability and lower cost.

Hemodynamic Characteristics of the Isolated Kidney

The perfusion flow produced by the pump is high with values up to 36 ml/min.[3] In our laboratory, we use flows close to 15 ml/min. The perfusion flow maintains the perfusion pressure of the renal artery. It is convenient that the perfusion pressure is in a rank of 90 ± 10 mmHg, although some authors obtain stable experimental models with higher perfusion pressure levels.[3] This pressure in the renal artery is controlled by monitoring the manometer and adjusting the pump flow. The pressure is measured on the end of the catheter that is inserted in the artery.

The high perfusion flow is due to the lower viscosity of the solution of perfusion with respect to the blood, and it is essential to maintain the adequate oxygenation of the renal tissue.[4]

Glomerular Filtration Rate of the Experimental Model

Generally, GFR values measured in an isolated kidney are lower than the ones found *in vivo*. The lower GFR leads to a decrease in the filtered load and, in some cases, to lower absolute reabsorption rates with respect to a kidney with normal GFR. Although this is not an important disadvantage of the experimental model, it must be taken into account when the obtained results are analyzed and interpreted.[1]

Other Characteristics

Sodium and water reabsorption, although with high and stable values, are not found at the same levels as in the intact rat, particularly if the relatively low GFR of the experimental model is considered. The defect would be located in the distal nephron.

Urine is almost isosmolar with respect to the perfusion, and the capacity of concentration and dilution is severely damaged. The fractional excretion of phosphate is low and remains stable for 2 hours when the studied kidney is perfused with normal concentrations of inorganic phosphate (1.5 mM). The apparently abnormal tubular handling of phosphate can be a consequence of the low filtrated loads. Nevertheless, a fact that cannot be dismissed is that the intracellular concentration of phosphate in the isolated kidney can be lower than what is normal, leading to an over-absorption of this electrolyte. However, the experimental model is adequate for studies related to the handling of phosphate, at least from the qualitative point of view.

Equipment

Figure 33.2 shows the arrangement of the elements used in the perfusion and the direction of the flow. As explained before, four catheters are introduced into the aorta artery, vein cava, femoral artery, and the ureter. The catheter placed in the aorta is connected to the manometer and to a chamber that works as a bubble trap. It also has a fork that takes the excess of buffer to the container with the solution of perfusion. In this way, once the perfusion pressure is stated, the renal flow will be adequate for the physiology of the organ and will not be fixed by the perfusion pump. The bubble trap has four connections: (1) an entrance to the buffer from the pump, (2) an exit to the aorta, (3) a manometer, and (4) an exit with a valve to allow the draining of the system. The peristaltic pump propels the buffer of perfusion, which is in the container. The container is thermostatized, mixed with a magnetic stirrer, and gassed with carbogen (95% O_2, 5% CO_2). Another catheter that takes the perfusate from the vein cava also goes to this container. A filter must be placed between the pump and the bubble trap to stop the particles bigger than 50 μm.

Catheters of adequate lengths are used to allow for comfortable working in the surgery field. However, they should not be excessively long because the buffer may reach the kidney without the adequate temperature. If the catheters are long, it is advisable to use a heating system near the

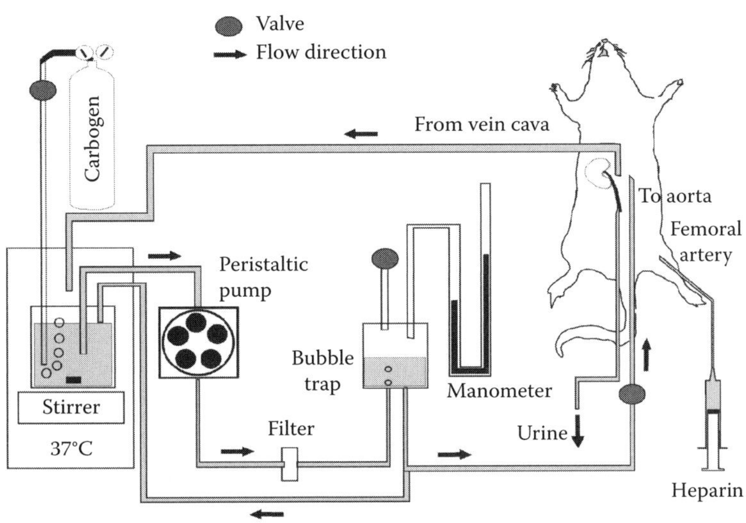

FIGURE 33.2 Shown is the relation of the different parts involved in isolated kidney perfusion.

entrance to the aorta. Placing two- or three-way valves in some parts of the system will make it easier to drain and clean.

The bubble trap, to which the manometer is connected, must be at the level of the entrance to the aorta. If it is not possible, it is important to remember that, per each centimeter above the animal, 0.75 mmHg must be added to the pressure measured in the manometer.

Viability of the Experimental Model

In order to evaluate the viability of the experimental model, different parameters are determined during perfusion. With this aim, samples of the solution of perfusion are extracted and urine is collected at different times during the experiment. Urinary flow is measured and the concentrations of the following substances are determined: inulin, glucose, and sodium. With the obtained values, the glomerular filtration rate and the fractional excretion of sodium and glucose are calculated. Also, the values of pressure and flow of perfusion must be monitored. All of these parameters must be constant during the experiment.

Experimental Results

The effect of sodium fluoride (NaF) on renal handling of phosphate was investigated.[6] Experiments were carried out in 50-day-old female inbred IIM/Fm strain subline "m" rats. Perfusion of the right kidney was performed using the buffer of perfusion with the composition and the conditions as indicated above. After a 20-minute stabilization period, four 5-minute urine collection periods were started and arterial perfusate samples were taken at midpoint. After that, a rectangular wave of 50 μM of NaF (n = 4) was produced in the buffer of perfusion for 30 minutes and urine and samples of the buffer were collected as stated above. Control rats (n = 3) were perfused under the same conditions without the addition of fluoride, to assess the viability of the kidneys throughout the perfusion.

Phosphate, glucose, inulin, and sodium were determined in buffer and urine samples. Urine volume was measured gravimetrically in preweighed collection vials.

The functional viability of the kidney was assessed by the glomerular filtration rate (estimated by inulin clearance), urine flow, and the percentage of tubular reabsorption of water ($TRH_2O\%$), sodium (TRNa%), and glucose (TRGlu%). Urinary phosphate was expressed as μg/min/g kidney weight.

As the values of functional parameters do not differ among collecting periods, the data were pooled in two periods: before and after fluoride addition to buffer of perfusion.

Table 33.1 displays functional parameters of an isolated rat kidney and it shows that no differences showed in any of the variables measured before and after the fluoride addition.

Figure 33.3 shows that urinary excretion of phosphate did not change throughout the experiment, analysis of variables (ANOVA), p >0.05.

TABLE 33.1
Functional Parameters of Isolated Rat Kidney

	PP (mmHg)	UF (μl/min)	GFR (ml/min/g Kidney Weight)	$TR_{Na}\%$ (Cl_{Na}/GFR)	$TR_{Glu}\%$ (Cl_{Glu}/GFR)	$TR_{H2O}\%$ (UF/GFR)
Without NaF	95 ± 5	51.6 ± 5.9	116.3 ± 12.2	61.2 ± 2.8	89.3 ± 3.2	61.1 ± 3.8
With F	95 ± 5	45.8 ± 6.5	108.8 ± 11.9	61.9 ± 2.7	90.0 ± 1.5	61.4 ± 2.4

Note: The results are expressed as the mean ± SEM, n = 4, PP: perfusion pressure, UF: urinary flow, GFR: glomerular filtration rate, ClNa: sodium clearance, ClGlu: glucose clearance.

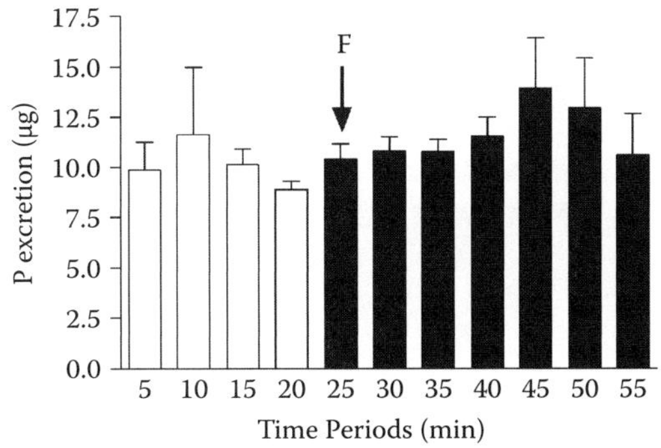

FIGURE 33.3 Urinary phosphate in isolated kidney is shown. Bars represent mean ± SEM of four experiments. F indicates the beginning of the square wave of NaF 50 μM.

REFERENCES

1. Maack, T. 1986. Renal clearance and isolated kidney perfusion techniques. *Kidn. Int.* 30: 142–151.
2. Shurek, H.J., and J.M. Alt. 1981. Effect of albumin on the function of perfused rat kidney. *Am. J. Physiol.* 240(9):.F569–F576.
3. Wang, J., R.L. Nation, A.M. Evans, and S. Cox. 2004. Isolated rat kidney perfused with dextran and bovine serum albumin: A stable model for investigating renal drug handling. *J. Pharmacol. Toxicol. Meth.* 49: 105–113.
4. Bekersky, I. 1983. Use of the isolated perfused kidney as a tool in drug disposition studies. *Drug Metab. Rev.* 14(5): 931–969.
5. Epstein, F.H., J.T. Prosnan, J.D. Tange, and B.D. Ross. 1982. Improved function with amino acids in the isolated perfused kidney. *Am. J. Physiol.* 243(12): F284–F292.
6. Di Loreto, V., A. Rigalli, and R.C. Puche. 2006. Effect of sodium fluoride administration to rats on bone phosphorous content and phosphatemia. *Arzneimittelforschung* 56:760–766.

34 Acute Renal Failure Models

Gisela Di Giusto and Adriana M. Torres

INTRODUCTION

Acute renal failure (ARF) is a syndrome characterized by a rapid (hours to weeks) decline in the glomerular filtration rate and retention of nitrogenous waste products, such as blood urea nitrogen and creatinine.

The diagnosis of ARF usually hinges on serial analysis of blood urea nitrogen and serum creatinine and, despite some limitations, these measurements are likely to remain the principal method for detection of ARF.

ARF is a combination of diseases that, for purposes of diagnosis and management, are conveniently divided into three categories:

1. Prerenal ARF, includes diseases characterized by renal hypoperfusion, in which the integrity of renal parenchymal tissue is preserved.
2. Intrinsic renal ARF, includes diseases involving renal parenchymal tissue.
3 Postrenal ARF, involves diseases associated with acute obstruction of the urinary tract.

Most acute intrinsic renal failure is caused by ischemia or nephrotoxins. Prerenal ARF and ischemic ARF are part of a spectrum of manifestations of renal hypoperfusion. Prerenal ARF is a response to mild or moderate hypoperfusion and ischemic ARF is the result of more severe or prolonged hypoperfusion, and both often coexistent with other renal trauma. They differ in that ischemic ARF, unlike prerenal ARF, is associated with injury to renal parenchyma and does not resolve immediately on restoration of renal perfusion. In its extreme form, renal hypoperfusion may result in bilateral renal cortical necrosis and irreversible renal failure. Ischemic ARF is observed most frequently in patients who have major surgery, trauma, severe hypovolemia, overwhelming sepsis, and burns.[1]

Urinary tract obstruction is a serious and common clinical condition, which is associated with increased intraluminal pressure in the ureter and renal tubules that may cause renal parenchymal damage through a series of direct and indirect effects.[1-3] Ureteral obstruction is a disorder that generally affects males with urinary retention due to hypertrophy of the prostate and patients with neurogenic bladder disorder.

UTILITY

The current concepts regarding the mechanisms of the impaired renal function and cellular injury in ARF are derived, predominantly, from studies in experimental animals and *in vitro* systems, such as isolated perfused kidneys, isolated proximal tubules in suspension, and renal cells in culture. The value of these models lies in defining the multiple factors that may potentially contribute to the pathophysiology of renal injury. However, it is important to appreciate that no experimental model faithfully replicates the process of renal injury in humans and the conclusions drawn from these studies should be extrapolated to the pathophysiology of human ARF with appropriate caution.[1-3]

MODEL OF ISCHEMIC ACUTE RENAL FAILURE

Studies of experimental animals have proved useful in documenting the temporal sequence of morphologic injury in ischemic ARF and defining the functional consequences of these changes at the whole kidney, single nephron, and cell level.

Renal ischemia induced by total cessation of renal blood flow is the most extensively studied experimental model of ischemic ARF.[1] The bilateral ischemia-induced ARF model consists in the occlusion of both renal pedicles with a smooth surface vascular clamp. This is a well-characterized model with some advantages:

- It is possible to make observations during both nonoliguric and oliguric phases of ischemia-induced ARF.
- Obviates the problem that exists with the unilateral ischemia with a contralateral nephrectomy model, that there might be adaptive changes in epithelial morphology, glomerular perfusion, and filtration rates due to the nephrectomy alone.
- Shares a spectrum of clinical features of human ARF.[4]

Materials

General anesthetic
Thermostatized rat board
Ring-handled preparation scissors
Ring-handled iris scissors
Cotton pliers (2)
Adson forceps
Artery forceps (2)
Rat-tooth forceps
Needle holder
Spatula
Nontraumatic vascular clamp (2)
Iodopovidone
Injection needle 15 × 0.5
Saline solution
Heparin
Suture needle ½ – 5 to 10 mm
Suture thread
Cotton buds

Procedure

Animals are anesthetized with sodium thiopental (70 mg/kg body weight, i.p.) and placed on a thermostatized rat board so as to maintain a constant body temperature. Both kidneys are exposed through flank incisions, mobilized by being dissected free from the perirenal fat. In order to induce the experimental acute renal failure, both renal pedicles are occluded with a smooth surface vascular clamp for 60 minutes (Figure 34.1).

Total ischemia is confirmed by observing blackening of the entire kidney surface. During the period of ischemia, the wound should be closed temporarily to maintain body temperature. After this ischemic period, the clamps are removed and reperfusion is allowed. The kidneys are observed for an additional 2 to 5 minutes to ensure color change, indicating blood reflow, and the wound is closed with 3-0 silk and surgical metal clamps. Experimental studies are performed after different times of reflow. Plasma urea levels are determined in order to corroborate the onset of the experimental pathology.[5]

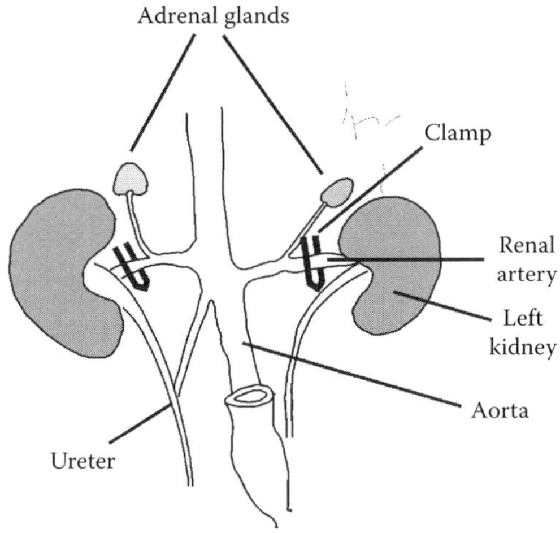

FIGURE 34.1 Illustrated is a model of ischemic acute renal failure showing the location of the nontraumatic vascular clamps.

MODEL OF OBSTRUCTIVE NEPHROPATHY

Although obstructive uropathy in humans is often partial and of long duration, most studies of the mechanisms of renal dysfunction in this condition have used models of complete obstruction for relatively short periods, such as 24 hours. These models avoid such issues as the degree of obstruction and the effects of long-term changes in renal architecture brought on by fibrosis or inflammation. Complete obstruction of short duration results in profound alterations in renal hemodynamics and glomerular filtration, as well as in tubule function, with minimal anatomic changes. Thus, acute complete obstruction provides an excellent model of regulation of renal function. The models of unilateral ureteral obstruction and bilateral ureteral obstruction have been employed.[2,3] We used the latter one in our laboratory.[6,7]

Materials

General anesthetic
Thermostatized rat board
Ring-handled preparation scissors
Ring-handled iris scissors
Cotton pliers (2)
Adson forceps
Artery forceps (2)
Rat-tooth forceps
Needle holder
Spatula
Nontraumatic vascular clamp
Injection needle 15 × 0.8
Saline solution
Heparin
Iodopovidone
Suture needle ½ – 5 to 10 mm
Suture thread
Cotton buds

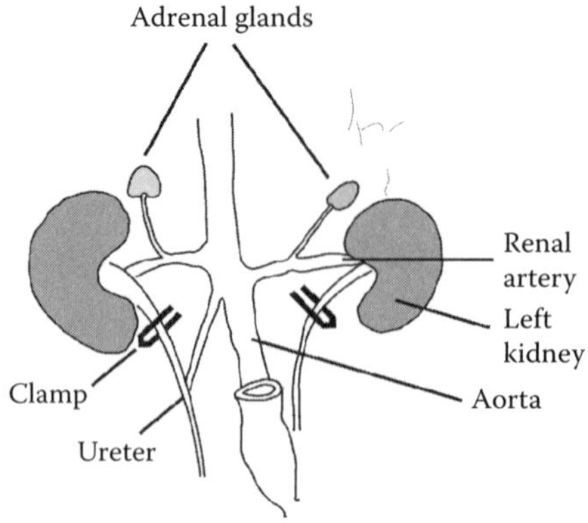

FIG. 34.2 Shown is a model of the obstructive nephropathy illustrating the location of the nontraumatic clamp.

Procedure

Animals are anesthetized and placed on a thermostatized rat board so as to maintain a constant body temperature. The abdominal cavity is opened, and a nontraumatic clamp is placed on both proximal ureters (Figure 34.2). After the closure of the abdomen, the animals are kept alive for 24 hours while they are given food and water *ad libitum*. The rats are considered to have a successful ureteral obstruction when the ureteral diameter is greater than 2 mm and hydronephrosis is evident. The ureteral obstruction is released after 24 hours (t = 0) and the rats, referred to as bilateral ureteral obstruction (BUO) treated rats, are kept alive and studied after different periods of time.[6,7]

Experimental Results

Some results obtained in our laboratory to characterize these experimental models are briefly described below and shown in Table 34.1 and Table 34.2.

Male Wistar rats aged from 110 to 130 days old were used throughout the study. Blood samples were withdrawn from the tail artery in order to measure urea plasma levels. The plasma was separated by centrifugation. These samples were used to measure urea plasma levels employing a commercial kit (Wiener Lab, Rosario, Argentina).

TABLE 34.1
Body Weight, Kidney Weight, Kidney/Body Weight Ratio, Urea Plasma Levels, and Cortical Renal Blood Flow (cRBF) in Sham (n = 4) and in Ischemic Acute Renal Failure (ARF, n = 4) Rats

	Sham	ARF
Body weight (g)	347.4 ± 2.4	345.8 ± 3.2
Kidney weight (g)	2.56 ± 0.11	3.09 ± 0.14*
Kidney/ body weight	0.0074 ± 0.0004	0.0089 ± 0.0004*
Urea plasma levels (g/l)	0.49 ± 0.02	0.58 ± 0.03*
cRBF (ml/min/100 g)	5.10 ± 0.55	3.64 ± 0.19*

Note: The results are expressed as mean ± SEM. Statistical analysis was performed using an unpaired Student's t-test, * $p < 0.05$.

TABLE 34.2
Body Weight, Kidney Weight, Kidney/Body Weight Ratio, Urea Plasma Levels, and Cortical Renal Blood Flow (cRBF) in Sham (n = 6) Rats and in Rats with Bilateral Ureteral Obstruction After 1 Day (BUO-1, n = 5) and After 2 Days of Ureteral Release (BUO-2, n = 6)

	Sham	BUO-1	BUO-2
Body weight (g)	327 ± 6	326 ± 10	329 ± 5
Kidney weight (g)	2.26 ± 0.07	3.22 ± 0.12[a]	3.43 ± 0.09[a]
Kidney/body weight	0.0070 ± 0.0002	0.0099 ± 0.0003[a]	0.010 ± 0.0003 [a]
Urea plasma levels (g/l)	0.56 ± 0.03	2.91 ± 0.30[a,c]	1.78 ± 0.25[a,b]
cRBF (ml/min/100 g)	6.44 ± 0.61	5.15 ± 1.28[c]	2.20 ± 0.35[a,b]

[a] $p < 0.05$ versus Sham, [b] $p < 0.05$ versus BUO-1, [c] $p < 0.05$ versus BUO-2.

Note: Results are expressed as mean ± SEM. Statistical analysis was performed using analysis of variance and Newman–Keuls posttest.

Cortical renal blood flow (cRBF) was determined in different sets of experimental animals using fluorescent microspheres (see Chapter 35 for explanation).

Ischemic Acute Renal Failure Model

All the studies of renal function were performed after 60 minutes of reperfusion (ARF group, n = 4). As a control group, rats were subjected to a sham operation identical to the one used for ARF rats without occlusion of both renal pedicles (Sham group, n = 4). These animals were monitored in parallel with ARF rats.

Body weight, kidney weight, kidney/body weight ratio, urea plasma levels, and cRBF are shown in Table 34.1. Kidney weight, kidney/body weight ratio, and urea plasma levels are increased significantly in the ARF group in comparison to the Sham rats. The cRBF is diminished in ARF animals in comparison with Sham ones.[5,8]

Obstructive Nephropathy Model

After 1 or 2 days of ureteral releasing (BUO group), the rats were anesthetized with sodium thiopental (70 mg/kg body weight, intraperitoneal (i.p.)) and all the measurements were recorded. Corresponding control groups, referred as Sham groups, were treated in the same way, except that no ureteral obstruction was performed.

Body weight, kidney weight, kidney/body weight ratio, urea plasma levels, and cRBF are shown in Table 34.2. After 1 or 2 days of ureteral release, a significant increase was observed in kidney weight, kidney/body weight ratio, and urea plasma levels as compared with Sham rats. cRBF only decreased after 2 days of ureteral releasing.[6,7]

REFERENCES

1. Brady, H.R., B.M. Brenner, and W. Lieberthal. 1996. Acute renal failure. In *The Kidney*, 5th ed., eds. B.B. Brenner and F.C. Rector. Philadelphia: WB Saunders Company, pp. 1200–1252.
2. Klahr, S. 2000. Obstructive uropathy. In *The kidney physiology and pathophysiology*, 3rd ed., eds. D.W. Seldin and G. Giebisch. Philadelpha: Lippincott Williams & Wilkins, pp. 2473–2512.
3. Curhan, G.C., and M.L. Zeidel. 1996. Urinary tract obstruction. In *The Kidney*, 5th ed., eds. B.B. Brenner and F.C. Rector. Philadelphia: WB Saunders Company, pp. 1200–1252.

4. Kwon, T.H., J. Frokiaer, P. Fernández-Llama, M.A. Knepper, and S, Nielsen. 1999. Reduced abundance of aquaporins in rats with bilateral ischemia-induced acute renal failure: Prevention by α-MSH. *Am. J. Physiol.* 277: F413–F427.
5. Di Giusto, G. 2006. Eliminación de aniones orgánicos en ratas con isquemia renal bilateral. Trabajo de Tesina para optar al título de Licenciado en Biotecnología, pp 22–24, 37–38, 53.
6. Villar, S.R., A. Brandoni, N.B. Quaglia, and A.M. Torres. 2004. Renal elimination of organic anions in rats with bilateral ureteral obstruction. *Biochim. Biophys. Acta.* 1688: 204–209.
7. Villar, S.R., A. Brandoni, N. Anzai, H. Endou, and A.M. Torres. 2005. Altered expression of rat renal cortical OAT1 and OAT3 in response to bilateral ureteral obstruction. *Kidn. Int.* 68: 2704–2713.
8. Di Giusto, G., N. Anzai, H. Endou, and A.M. Torres. 2008. Elimination of organic anions in response to an early stage of renal ischemia-reperfusion in the rat: Role of basolateral plasma membrane transporters and cortical renal blood flow. *Pharmacology* 81: 126–137.

35 Renal Blood Flow Measurement

Gisela Di Giusto, Adriana M. Torres, and Alfredo Rigalli

RENAL BLOOD FLOW MEASUREMENT BY HYDROGEN DESATURATION

Utility

This technique is useful not only for renal blood flow measurement, but also for the measurement of blood flow in other organs or tissues. This technique establishes a comparison of renal blood flow between organs under the same experimental conditions or in the same organ under different experimental conditions.

Materials

General anesthetic
Thermostatized rat board
Ring-handled preparation scissors
Ring-handled iris scissors
Adson forceps (2)
Artery forceps (2)
Rat-tooth forceps
Needle holder
Three-way valve
Voltmeter
Platinum electrode
Reference electrode
Hydrogen
Catheter for tracheostomy
Iodopovidone
Suture needle curved ½– 5 to 10 mm
Suture thread
Cotton buds

Procedure

The tracheostomy is performed under general anesthesia. The endotracheal catheter is connected to a three-way valve, which allows for administering air or a mixture of hydrogen and air to the animal. Once the tracheostomy is performed, the animal is placed in a ventral decubitus position and a paramedial incision is performed in the lumbar region. The retroperitoneal fat pad is exposed and the kidney is easily visualized. After exposing the kidney, a 3-mm-long platinum electrode is implanted in the kidney (cortex or medulla). The electrode must have a diameter of 0.05 mm. An Ag/AgCl (silver/silver chloride) reference electrode is placed in a container with KCl (potassium chloride) solution, which must be saturated with AgCl. The container has a catheter with a 15 ×

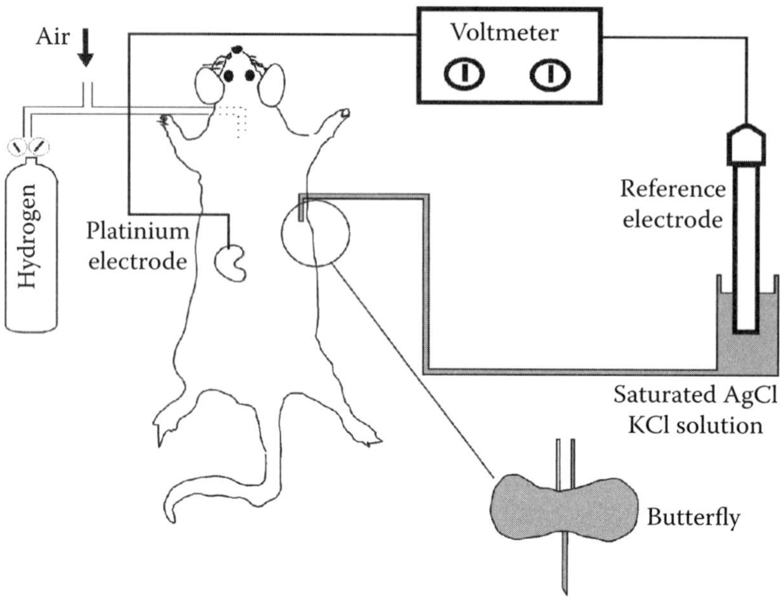

FIGURE 35.1 Shown is scheme of materials for the measurement of renal blood flow by desaturation of hydrogen.

0.5 injection needle in the opposite end. This needle is subcutaneously inserted in the animal and acts as a salt bridge between the reference and the hydrogen electrode. The voltage developed by the hydrogen and the reference electrode is measured with a voltmeter and/or recorded with the appropriate device.

The animal is allowed to breathe air through the three-way valve and the voltage is recorded (basal voltage). After a stabilization period, usually 5 to 10 minutes, the valve is changed and the animal is supplied with a mixture of hydrogen and air. This condition is maintained until the voltage reaches a constant value (plateau). After that, air is supplied to the animal and voltage is recorded until it reaches the basal voltage. The time needed to reach the middle value of voltage between the plateau and the basal value is called *half-life of desaturation*. This time is inversely proportional to renal blood in the region where the electrode is placed.[1] The relationship between the electrode, the animal, and the device for hydrogen supply is displayed in Figure 35.1

Precaution

Hydrogen is odorless and highly explosive. As a consequence, the catheters and connectors must be carefully checked in order to recognize the leakage of hydrogen. This procedure must be done with a sponge and water with detergent. Bubble formation indicates a leak of hydrogen. Never use a flame to check the hydrogen leakage. The experiment must be carried out in a well-ventilated laboratory.

Experimental Results

Cortical renal blood flow (cRBF) was measured in four rats treated with 80 μmol of sodium fluoride (NaF) by an orogastric catheter. Four animals were selected to receive water by an orogastric catheter and to act as controls. There was no difference in half-lives of desaturation between the groups. Controls included: 44.10 ± 2.89 min, NaF-treated: 44.14 ± 2.80 min, $p > 0.05$.

RENAL BLOOD FLOW MEASUREMENT WITH FLUORESCENT MICROSPHERES

UTILITY

Regional organ perfusion can be estimated with hematogenously delivered microspheres.[2] When appropriately sized microspheres are used, regional blood flow is proportional to the number of microspheres trapped in the region of interest.[3]

Since introduced by Rudolf and Heyman[4] in 1967, measurement of regional organ blood flow using radiolabeled microspheres has evolved to become the gold standard. However, there are increasing concerns regarding health and environmental hazards, and the expense associated with special handling, disposal, and limited shelf-life.

Techniques using nonradioactive colored or fluorescent microspheres have been developed and validated against traditional methods.[5] In particular, the fluorescent microspheres have a shelf-life of at least 1 year. They also offer higher sensitivity, superior color separation, and greater ease of measurement compared to ones that are colored.

A number of studies showing the decrease in renal blood flow (RBF) in ischemic acute renal failure (ARF) evaluated this parameter using the renal clearance of p-aminohippurate (PAH). In this regard, it is important to mention that almost 70 years have elapsed since Homer Smith first postulated that PAH is an ideal marker to determine the rate of renal plasma flow. Provided that plasma level was kept low enough to prevent saturation of its tubule cell transporter, PAH could be shown to be almost completely extracted from the renal circulation and excreted in the urine.[6] Since then, Smith's proposal has been validated in subjects with healthy kidneys in which the extraction ratio of PAH repeatedly has been shown to approximate 0.9.[7] Smith's assumption does not appear to be valid in the presence of postischemic acute renal failure. Under the latter circumstances, the extraction ratio of PAH has been shown to be severely impaired, with the result that PAH clearance markedly underestimates renal plasma flow.[8] Corrigan et al.[9] have determined the effect of postischemic injury to the human renal allograft on PAH extraction and renal blood flow. They demonstrated that an ensuing reduction in urinary PAH clearance results in a gross underestimation of renal plasma flow, which is close to the normal range in the initiation, maintenance, and recovery stages of an injury. In agreement with this, we have demonstrated that urinary PAH clearance is not in correlation with renal plasma flow measured with fluorescent microspheres in different experimental pathological models, such as bilateral ureteral obstruction[10] and acute biliary obstruction.[11]

Cortical renal blood flow (cRBF) is determined using fluorescent microspheres because clearance of PAH is not an adequate estimation of cRBF in postischemic ARF, as was described by Corrigan et al.[9]

Materials

General anesthetic
Thermostatized rat board
Ring-handled preparation scissors
Cotton pliers (2)
Adson forceps
Artery forceps
Rat-tooth forceps
Spatula
Withdrawal pump
Polycarbonate filters
Fluorescent microspheres
Fluorometer
2-(2-ethoxyethoxy) ethyl acetate

Catheter
Injection needle 25 × 0.8
Saline solution
Potassium hydroxide
Heparin
Suture thread
Cotton buds

Procedure

Animals are anesthetized with sodium thiopental (70 mg/kg body weight, intraperitoneal (i.p.)) and placed on a thermostatized rat board so as to maintain a constant body temperature. Then the carotid artery and the femoral artery are catheterized.

Orange microspheres (540 to 560 nm) with a diameter of 15 μm (molecular probes) are used to evaluate renal blood flow. They are infused (10^6/kg body weight) in bolus into the carotid artery followed by a washout with 1 ml of saline. Arterial reference blood is collected from the femoral artery at a rate of 1 ml/min starting 15 seconds before injection with a withdrawal pump and continuing for 60 seconds after injection. Five minutes after the injection, the rats are killed and the kidneys are rapidly removed and dissected to obtain medulla and cortex. Blood samples are digested overnight with 89.2% potassium hydroxide (KOH). The renal tissue is digested for 24 hours with 22.4% KOH. A known amount of yellow-green microspheres (505 to 515 nm) is added to each sample vial of blood and tissue prior to digestion. These microspheres act as an internal standard. Digested samples are filtered using Poretics polycarbonate filters. The filters are introduced in 2-(2-ethoxy-ethoxy) ethyl acetate and shaken for at least 1 hour in order to extract the fluorescent dye. Then the fluorescence is measured at 540 to 560 nm and at 505 to 515 nm.

The RBF (ml/min) is calculated with the formula:

$$\mathrm{RBF} = \frac{\mathrm{fl}}{\mathrm{fl}_{\mathrm{ref}}} \times \mathrm{R} \tag{35.1}$$

where fl is the fluorescence of the renal tissue, $\mathrm{fl}_{\mathrm{ref}}$ is the fluorescence of the reference blood sample, and R is the withdrawal rate of the reference blood sample (ml/min).

Precaution

Accurate measurement of regional organ perfusion requires that the microspheres are uniformly dispersed in the blood. Aggregation of the microspheres should be minimized by ultrasonicating the microsphere suspension prior to use and vigorously vortexing them prior to injection.

Experimental Results

The aim of the present study was to evaluate the cRBF in rats with ischemic ARF. We have described a decrease of glomerular filtration rate and of RBF in ischemic ARF.[12,13] Male Wistar rats aged from 110 to 130 days old were used throughout the study (340 to 355 g body weight). The rats were anesthetized as described previously. Through flank incisions, the renal pedicles were clamped for 60 minutes with a nontraumatic vascular clamp. After 60 minutes of reperfusion, the cRBF was determined (ARF group, n = 6). As a control group, rats were subjected to a Sham operation identical to the one used for ARF rats without occlusion of both renal pedicles (Sham group, n = 6). These animals were monitored in parallel with ARF rats. cRBF decreased more in ARF animals than in the Sham group (Table 35.1).[14]

TABLE 35.1
Cortical Renal Blood Flow (cRBF) in Rats with Acute Renal Failure (ARF) and in Sham Rats

	Sham	ARF
cRBF (ml/min/100 g body weight)	5.10 ± 0.55	3.64 ± 0.29*

* $p < 0.05$.
Note: Results are expressed as mean ± SEM.

REFERENCES

1. Haining, J.L., and M.D. Turner. 1966. Tissue blood flow in rat kidney by hydrogen desaturation. *J. Appl. Physiol.* 21(6): 1705–1708.
2. Heyman, M.A., B.D. Payne, J.I. Hoffman, and A.M. Rudolf. 1977. Blood flow measurements with radionuclide-labeled particles. *Prog. Cardiovasc. Dis.* 20: 55–79.
3. Bassingthwaighte, J.B., T.C. Malone, R.B. Moffett, R.B. King, I.S. Chan, J.M. Link, and K.A. Krohn. 1990. Molecular and particulate depositions for regional myocardial flows in sheep. *Circ. Res.* 66: 1328–1344.
4. Rudolph, A.M., and M.A. Heyman. 1967. The circulation of the fetus in utero. Methods for studying distribution of blood flow, cardiac output and organ blood flow. *Circ. Res.* 21: 163–184.
5. Glenny, R.W., S. Bernard, and M. Brinkley. 1993. Validation of fluorescent-labeled microspheres for measurement of regional organ perfusion. *J. Appl. Physiol.* 74: 2585–2597.
6. Smith, H.W., N. Finkelstein, L. Aliminosa, B. Crawford, and M. Graber. 1945. The renal clearances of substituted hippuric acid derivatives and other aromatic acids in dog and man. *J. Clin. Invest.* 24: 388–404.
7. Battilana, C., H. Zhang, R.A. Olshen, L. Wexler, and B.S. Myers. 1991. PAH extraction and estimation of plasma flow in diseased human kidneys. *Am. J. Physiol.* 261: F726–F733.
8. Myers, B.D., C. Miller, J. Mehigan, C. Olcott, H. Golbetz, C.R. Robertson, G. Derby, R, Spencer, and S. Friedman. 1984. Nature of the renal injury following total renal ischemia in man. *J. Clin. Invest.* 73: 329–341.
9. Corrigan, G., D. Ramaswamy, O. Kwon, F.G. Sommer, E. Alfrey, D.C. Dafpe, R.A. Olshen, J.D. Scandling, and B.D. Myers. 1999. PAH extraction and estimation of plasma flow in human post-ischemic acute renal failure. *Am. J. Physiol.* 277: F312–F318.
10. Villar, S.R., A. Brandoni, N.B. Quaglia, and A.M. Torres. 2004. Renal elimination of organic anions in rats with bilateral ureteral obstruction. *Biochim. Biophys. Acta.* 1688: 204–209.
11. Brandoni, A., N.B. Quaglia, and A.M. Torres. 2003. Compensation increase in organic anion excretion in rats with acute biliary obstruction: Role of the renal organic anion transporter 1. *Pharmacology* 68: 57–63.
12. Brady, H.R., B.M. Brenner, and W. Lieberthal. 1996. Acute renal failure. In *The Kidney*, 5th ed., eds. B.B. Brenner and F.C. Rector. Philadelphia: WB Saunders Company, pp. 1200–1252.
13. Green, J., Z. Abassi, J. Winaver, and K.L. Skorecki. 2000. Acute renal failure: Clinical and pathophysiologic aspects. In *The Kidney: Physiology and pathophysiology*, 3rd ed., eds. D.W. Seldin and G. Giebisch. Philadelphia: Lippincott, Williams & Wilkins, pp. 2329–2373.
14. Di Giusto, G., N. Anzai, H. Endou, and A.M. Torres. 2008. Elimination of organic anions in response to an early stage of renal ischemia-reperfusion in the rat: Role of basolateral plasma membrane transporters and cortical renal blood flow. *Pharmacology* 81: 126–137.

36 Transport Studies in Plasma Membrane Vesicles Isolated from Renal Cortex

Gisela Di Giusto and Adriana M. Torres

INTRODUCTION

Transcellular transport in renal tubules involves the movement of solutes between three compartments: luminal, intracellular, and peritubular. Two barriers separate them: the apical and the basolateral plasma membranes. For vectorial transport of solutes, a functional polarity of the cell is required, i.e., the transport properties of the apical and the basolateral membranes must be different. Indeed, the two membranes are different in almost every respect, such as morphology, enzyme content, protein–lipid–carbohydrate composition, hormone receptors, and transport properties. These differences are emphasized in the procedures for membrane separation. Different lipid-to-protein ratios and different carbohydrate contents result in different physical properties of the two membranes: density and surface charge, which allow for their separation by density gradient centrifugation, phase partitioning, differential precipitation, and electrophoresis.[1]

METHODS OF MEMBRANE ISOLATION

The selection of a method for membrane isolation is usually dictated by the aim of the study. For studies of kinetics, energetics, stoichiometry, biochemical identification, and reconstitution of a transport system, a fast isolation method providing high yield of sealed vesicles containing the intact transport system is of primary importance. Small cross contamination with other membranes may be acceptable, as long as the cross-contaminating structures do not contain high activities of the same or a similar transport system. For localization studies, on the other hand, a method that effectively separates all the membrane structures that might contain the transport system under study is mandatory; yield and membrane integrity might be of secondary importance. In general, one can distinguish between two kinds of methods: the analytical isolation procedures, which generate small amounts of purified membrane fractions, and preparative procedures, which allow fast separation of large amounts of membranes whereby purity is sacrificed for the sake of speed and yield.

Differences in enzyme activities provide the criteria for identification of the separated membranes. Na^+/K^+-ATPase is usually used as a marker for basolateral membranes and alkaline phosphatase for brush border membranes. However, one must be aware that marker enzymes might not be localized in one membrane exclusively, so a high enrichment factor does not exclude prior cross contamination.

The kidney is made up of highly heterogeneous tissue, and so are the membranes that are isolated. The heterogeneity can be reduced by careful anatomical separation of tissue zones before proceeding to membrane isolation. The outermost cortex contains mostly the proximal convoluted tubules of superficial nephrons, juxtamedullary cortex contains mainly the proximal tubules of deep nephrons, and the pars recta of the superficial nephrons and the outer medulla consists mostly of pars recta of both nephron populations. Anatomical preseparation of kidney tissue before membrane isolation allows the detection of heterogeneity of transport systems in different nephron segments.[1,2]

Utility

Isolation of vesicles formed by the brush border and the basolateral membranes made it possible to characterize a number of transport systems under well-defined *in vitro* conditions. By measuring solute flux into or out of the vesicles, the transport properties of the membrane can be determined. The use of membrane vesicles is a significant advantage over whole cells or models that use tissue instead of free cells or intact animals; most notable is the control over the supply of energy and the absence of complicating factors, such as intracellular metabolism or binding to intracellular structures. In addition, the complexity of transepithelial solute transport is reduced because of the physical separation of the two membranes, so that transport across one can be studied independently from the other. It is, however, important to refer the vesicle data back to the intact cell or tissue.

Materials

General anesthetic
Thermostatized rat board
Ring-handled preparation scissors
Ring-handled iris scissors
Cotton pliers (2)
Adson forceps
Artery forceps
Rat-tooth forceps
Potter homogenizer with Teflon pestle
5-ml syringe
Injection needle 25 × 0.8
Iodopovidone
Cotton buds

Procedure

Experimental Animals

Rats are anesthetized with sodium thiopental (70 mg/kg body weight, intraperitoneal (i.p.)) and the kidneys are removed. The renal tissue is blotted, weighed, and placed in a saline buffer. The cortex and medulla are separated and cortical tissue is used for the preparation of basolateral and brush border membrane vesicles.

Preparation of Basolateral Membrane Vesicles (BLMV) from Kidney Cortex

The preparations of BLMV are done by a modification of the method described by Jensen and Berndt[3] as previously reported by this chapter's authors.[4,5] Kidney cortical tissues are placed in a Dounce homogenizer containing 250 mM sucrose and 5 mM Hepes–Tris buffer (pH 7.40) at a ratio of 12.5 ml/g cortex wet weight. After four gentle strokes with a loose fitting pestle, the suspension is homogenized further with a motor-driven Teflon pestle and spun down for 15 minutes at 1200 × g. The supernatant is aspirated and spun for 15 minutes at 22,000 × g. The fluffy beige upper layer of the resulting pellet (crude plasma membranes) is resuspended in about 1 ml of supernatant and added to 19 ml of buffered sucrose. The membrane suspension is homogenized gently through a 16-gauge gavage needle followed by the addition of 3.7 ml of 100% Percoll™. The Percoll–membranes mixture is spun for 30 minutes at 39,250 × g. The top clear layer is discarded and the top-most dark band is removed. This layer is composed primarily of basolateral membranes. BLMVs are brought up in potassium chloride (KCl) buffer (85 mM KCl, 83 mM sucrose, 2 mM Hepes–Tris, pH 7.40) at a ratio of 8 ml/g original cortex wet weight. Next, BLMVs are pelleted at 30,000 × g for 30 minutes and washed three times with the KCl buffer. Finally, BLMVs are resuspended in 300 μl of 250 mM manitol and 10 mM Hepes–Tris buffer (pH 7.40). Aliquots of the membranes are stored immediately at –70°C until ready for use (no more than two weeks after membrane preparations).

Preparation of Brush Border Membrane Vesicles (BBMV) from Kidney Cortex

BBMVs are isolated from renal cortex by Mg^{2+}/EGTA precipitation as previously described by the authors.[4,5] The kidney cortex is cut off, minced, and homogenized in 30 vol. (v/w) of ice-cold, 50 mM sucrose, 2 mM Tris–HCl buffer (pH 7.10), and 5 mM EGTA for 5 minutes at top speed in a Sorvall Omnimixer. $MgCl_2$ is added to the homogenate for a final concentration of 12 mM and the mixture is stirred in an ice bath for 15 minutes. The homogenate is then centrifuged (3,000 × g, 15 min, 4°C). The supernatant is carefully decanted and centrifuged again at 28,000 × g for 40 minutes at 4°C. The pelleted material representing the brush border membranes is resuspended in 50 mM sucrose and 10 mM Hepes–Tris solution (pH 7.50) and centrifuged for 15 minutes at 800 × g at 4°C. The supernatant is finally centrifuged for 45 minutes at 28,000 × g. The brush border membrane pellets obtained are resuspended in 50 mM sucrose and 10 mM Hepes–Tris buffer (pH 7.40). Aliquots of the membranes are stored immediately at –70°C until used (no more than two weeks after membrane preparations).

Marker Enzyme Assays

Samples for enzymes and protein assays are removed from the homogenates and from the resuspended membranes. Alkaline phosphatase, as a marker for brush border membranes, is determined using a commercial kit (Wiener Lab, Rosario, Argentina). Na^+-K^+-ATPase activity, as a marker for basolateral membranes, is estimated as the difference between the amounts of inorganic phosphate liberated in the absence (total ATPase) and in the presence (Mg^{2+}-ATPase) of ouabain. The release of inorganic phosphate is measured according to the method described by Widnell.[6] Protein quantification of samples was performed using the method of Sedmak and Grossberg.[7]

Transport Studies of p-Aminohippurate (PAH)

PAH uptake by BBMV and BLMV is measured by the rapid filtration technique using a Millipore vacuum filtration system and nitrocellulose membranes with a pore size of 0.45 μm as previously reported.[4] All filters are soaked and presaturated overnight in 1 mM unlabelled PAH in 100 mM NaCl, 1 mM Hepes–Tris (pH 7.40) to minimize nonspecific binding of radioactivity. The degree of nonspecific binding to the filter is about 0.02%. Vesicles are rapidly thawed at 37°C and used for uptake studies.

BBMVs are preincubated in 100 mM sucrose, 100 mM K^+-gluconate, and 10 mM Hepes–Tris (pH 7.40) at 37°C for 1 hour. The reaction is started by adding 25 μl of vesicles (150 to 200 μg protein) to 300 μl of incubation medium at 37°C. The incubation medium contains 100 mM sucrose, 100 mM K^+-gluconate, 10 mM Hepes–Tris (pH 7.40), and 10 μCi ^{3}H-PAH to which unlabelled PAH is added to yield final concentration.

BLMVs (25 μl, 150 to -250 μg protein) are added to an incubation medium containing 100 mM NaCl, 100 mM sucrose, 20 mM Hepes–Tris (pH 7.40), 10 μM α-ketoglutarate, and 10 μCi ^{3}H-PAH to which unlabelled PAH is added to yield final concentration.

After a predetermined incubation time, the uptake is terminated both in BBMVs and BLMVs by diluting the samples to 1 ml with ice-cold stop solution (150 mM KCl, 20 mM Hepes–Tris, pH 7.40). The diluted samples are immediately filtered, and then the filters are washed three times with 3 ml of ice-cold stop solution. Radioactivity in the filters is measured in a liquid scintillation spectrometer counter. Uptake at time zero is assessed by simultaneously mixing 1 ml of ice-cold stop solution with 25 μl of vesicles and 300 μl of the isotope-containing buffer. The diluted sample is immediately filtered and washed three times with 3 ml of ice-cold stop solution. The uptake at time zero is subtracted from that obtained at the same PAH concentration after incubating at 37°C during a predetermined incubation time. Kinetic data obtained is fitted by the Michaelis–Menten model with a nonlinear, least square criterion using the Enzfitter program (Elsevier Biosoft).[4]

Experimental Results

Male and female Wistar rats aged from 110 to 130 days old were used throughout the study. Male and females weights (g) were 375 ± 12 and 234 ± 3, respectively. Animals were allowed free access to standard laboratory chow and tap water, and housed in a constant temperature and humidity environment with regular light cycles (12 hours). Kidneys were removed, BBMV and BLMV were prepared and PAH uptake studies were performed as described above.

Membrane Purification

Table 36.1 summarizes the specific activities and enrichment of marker enzymes in homogenates and final BBMV and BLMV isolated from male and female rats. The purification of renal brush border and basolateral membranes has been demonstrated by the enhancement of the specific activity of alkaline phosphatase in BBMV and that of Na^+-K^+-ATPase-specific activity in BLMV relative to those of their respective homogenate fractions. The specific activity of alkaline phosphatase in BBMV increased more than 12 times relative to that in their respective homogenate. The specific activity of Na^+-K^+-ATPase in BLMV was more than 8 times higher than that in their respective homogenate. Both membrane fractions were isolated from male and female rat kidneys with equal specific activities and enrichments of Na^+-K^+-ATPase and alkaline phosphatase. The enrichment of marker enzymes, both in BBMV and BLMV, were similar to those described for comparable techniques.[1-5]

TABLE 36.1
Specific Activities and Enrichment of Marker Enzymes in Rat Renal Cortex Homogenates, Brush Border Membrane Vesicles, and Basolateral Membrane Vesicles Isolated From Male (M) and Female (F) Rat Kidneys

	Brush Border Membrane		
	Homogenates	Vesicles	Enrichment
Na^+-K^+-ATPase	M: 9.74 ± 1.49	M: 5.33 ± 0.63	M: 0.55 ± 0.04
(μmol Pi/mg Prot./h)	F: 8.81 ± 1.07	F: 6.38 ± 2.19	F: 0.68 ± 0.19
Alkaline Phosphatase	M: 302 ± 25	M: 3975 ± 217	M: 13.28 ± 0.87
(mIU/mg Prot.)	F: 278 ± 38	F: 3373 ± 316	F: 12.30 ± 0.76
	Basolateral Membrane		
	Homogenates	Vesicles	Enrichment
Na^+-K^+-ATPase	M: 8.01 ± 1.10	M: 75 ± 10	M: 9.35 ± 0.75
(μmol Pi/mg Prot./h)	F: 10.14 ± 1.51	F: 81 ± 23	F: 8.44 ± 2.65
Alkaline Phosphatase	M: 307 ± 29	M: 763 ± 58	M: 2.49 ± 0.09
(mIU/mg Prot.)	F: 300 ± 21	F: 743 ± 42	F: 2.48 ± 0.16

Note: Results are expressed as mean ± SEM from experiments carried out in triplicate in three different vesicle preparations for each experimental group.

PAH Uptake Assays

The uptake of PAH in BBMV and BLMV has been determined in male and female rats. The time course of ^{3}H-PAH uptake by BBMV and BLMV was determined over a period of 30 minutes. Uptake velocity was linear up to 15 seconds in all experimental groups (data not shown). Accordingly, unless otherwise indicated, the 15 second time was chosen to measure PAH uptake to fulfill the requirements for measuring linear transport.

To elucidate whether the effect of the animal's sex is due to a variation in the Michaelis–Menten constant (Km) and/or the maximum velocity (Vmax) of PAH transport, the initial rate (15 s) of PAH uptake was measured as a function of the external PAH concentration as shown in Figure 36.1.

Kinetic parameters of PAH uptake in male BBMVs and BLMVs were similar to values previously reported in plasma membrane experimental models.[4,8] As can be seen in Table 36.2, PAH had the following difference in transport parameters due to the animal's sex:

1. BBMV: no variation was observed for apparent Vmax values, whereas an apparent Km value was significantly lower in males than in females. This indicates that the affinity of the PAH transporter located at BBMV is greater for males than for females.[5]

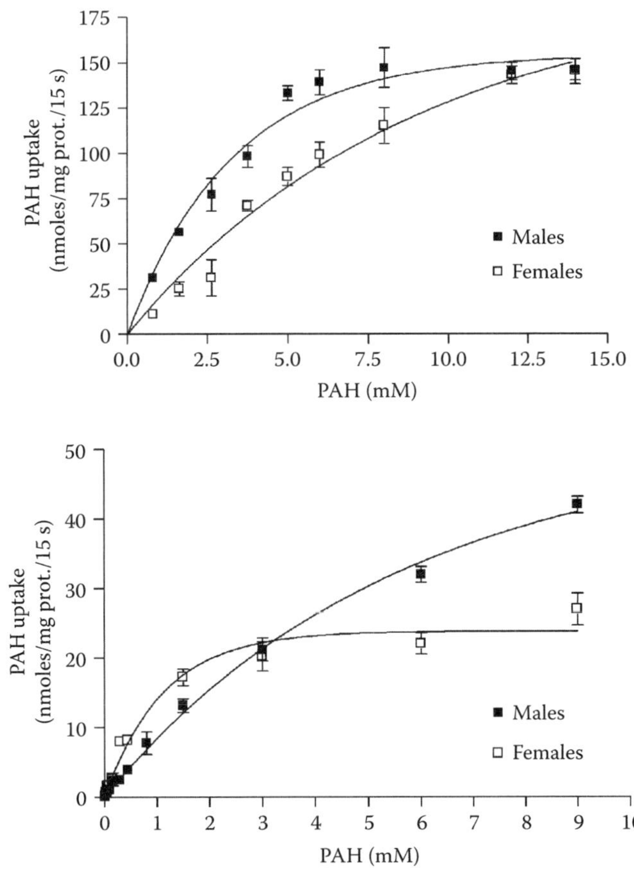

FIGURE 36.1 PAH uptake in renal brush border (upper panel) and basolateral membrane (lower panel) vesicles from male and female rats (nmoles/mg prot./15 s). Results are expressed as mean ± SEM from experiments carried out in triplicate in three different membrane preparations for each experimental group.

TABLE 36.2
Michaelis–Menten Kinetic Parameters of PAH Uptake in Brush Border Membrane Vesicles and in Basolateral Membrane Vesicles Isolated from Male and Female Rat Kidneys

	Vmax, (nmol/mg Prot./s)		Km, (mM)	
Membranes Obtained from	**Males**	**Females**	**Males**	**Females**
Brush border membrane vesicles	14.1 ± 1.1	15.3 ± 0.5	4.8 ± 0.3	7.8 ± 0.4*
Basolateral membrane vesicles	3.6 ± 0.1	1.9 ± 0.3*	4.6 ± 0.4	1.1 ± 0.2*

* $p < 0.05$.

Note: Results are expressed as mean ± SEM from experiments carried out in triplicate in three different membrane preparations for each experimental group.

2. BLMV: both Vmax and Km values were distinctly higher in male rats that in female rats. The difference in Vmax suggests that a higher number of functional carrier units exist in male BLMV, whereas the higher Km indicates that basolateral male carriers have a lower affinity.[5]

REFERENCES

1. Murer, H., and P. Gmaj. 1986. Transport studies in plasma membrane vesicles isolated from renal cortex. *Kidney Int.* 30: 171–186.
2. Sachs, G., R.J. Jackson, and E.C. Rabon. 1980. Use of plasma membrane vesicles. *Am. J. Physiol.* 238: 6151–6164.
3. Jensen, R.E., and W.O. Berndt. 1988. Epinephrine and norepinephrine enhance p-aminohippurate transport into basolateral membrane vesicles. *J. Pharmacol. Exp. Ther.* 244: 543–549.
4. Montagna, G., C.G. Hofer, and A.M. Torres. 1998. Impairment of cellular redox status and membrane protein activities in kidneys from rats with ischemic acute renal failure. *Biochim. Biophys. Acta.* 1407: 99–108.
5. Cerrutti, J.A., N.B. Quaglia, and A.M. Torres. 2001. Characterization of the mechanisms involved in the gender differences in p-aminohippurate renal elimination in rats. *Can. J. Physiol. Pharmacol.* 79: 805–813.
6. Widnell, C.B. 1974. Purification of rat liver 5' nucleotidase as a complex with sphingomyelin. *Meth. Enzymol.* 32: 368–374.
7. Sedmak, J.J., and S.E. Grossberg. 1977. A rapid, sensitive and versatile assay for protein using Coomassie brilliant blue G250. *Anal. Biochem.* 79: 544–552.
8. Chakrabarti, S., D. Due Vu, and M. Coté. 1991. Effects of cysteine derivatives of styrene on the transport of p-aminohippurate ion in renal plasma membrane vesicles. *Arch. Toxicol.* 65: 366–372.

37 Remnant Kidney Model

Maria V. Arcidiacono, Luis A. Ramirez, Valeria Dalmau, and Alfredo Rigalli

INTRODUCTION

Chronic renal failure in human beings is caused by kidney damage or when the glomerular filtration rate is less than 60 ml/min for 3 months or more. This is a progressive process that results in renal disease. Clinical studies show that high morbidity and mortality rates occur in patients suffering chronic kidney disease.

In patients with chronic renal failure, loss of renal function leads to various metabolic disturbances. These disturbances promote the development of secondary hyperparathyroidism and bone disease. Other symptoms include cardiovascular, neurological, hematological, gastrointestinal, and metabolic–endocrine disorders. Renal function declines progressively once the clearance of creatinine decreases 25% from its normal value. However, symptoms are often not apparent until renal failure is well advanced.

The remnant kidney model also known as 5/6 nephrectomy (5/6NX) is the most suitable animal model to investigate the mechanisms underlying chronic renal failure and its complications. This technique involves surgical removal of one kidney and the obstruction of two-thirds of the blood flow of the other kidney.[1]

MATERIALS

General anesthetic
Thermostatized rat board
Ring-handled preparation scissors
Ring-handled iris scissors
Cotton pliers (2)
Adson forceps
Artery forceps (2)
Rat-tooth forceps
Microforceps (2)
Needle holder
Stereomicroscope
Antibiotic
Analgesic
Suture needle ½ – 5 to 10 mm
Suture thread
Cotton
Cotton buds (10)
Gauze
Iodopovidone

Procedure

Once the rat is anesthetized, the abdominal area is shaved from the sternum to the pelvic area. The entire area is sterilized with iodopovidone. Using toothed forceps and scissors, 2 cm of the skin is cut from 1 cm below the sternum (for a 200 g rat). The abdominal muscles are then exposed and raised with forceps to prevent the organs from being damaged. The incision is then surrounded by gauze.

Another possibility is to make a 1-cm right paramedian lumbar incision to expose the right kidney, and a 1-cm left paramedian ventral incision for the left kidney (Figure 37.1). This procedure is particularly useful if surgeries of the left and right kidneys are made in two stages. However, the procedure with one incision is useful when surgeries of both kidneys are done on the same day.

Using a cotton swab, the intestine and the surrounding tissue are gently moved to the right to display the spleen, which is the reference point in finding the left kidney. Using the cotton swab, the left kidney is exposed and placed on the gauze. With the tip of the cotton swab, the capsule is removed from the pedicle using a delicate counterclockwise movement, and paying attention to the ureter. In the pedicle, the branches of the left renal artery (usually three in the front of the kidney) are visible and appear thinner than the vein (purple color), and are red with a white and transparent contour. Using a microforceps, two branches are raised slightly and, with the other forceps, are isolated from the connective tissue. This operation is risky because it is very easy to pierce the vein, causing bleeding. After lifting the branch, a 4.0 silk suture is passed around it. Then, the branch is tied off with two knots.

The left kidney is turned over and the remaining branches are isolated and tied to obstruct blood flow to two-thirds of the kidney. The kidney is then placed back into the abdominal cavity.

Next, the right kidney is exposed and the capsule is removed. The renal pelvis and the ureter are clamped with hemostats and a 2.0 silk suture floss is used for tying. Two knots are tied on one side and two on the other before the kidney is cut out. The vascular stump is placed back into the abdomen (Figure 37.2).

The remnant left kidney should be rechecked to ensure that the proper blood flow reduction was achieved. In case more than one-third of the kidney still has blood flow, more arteries should

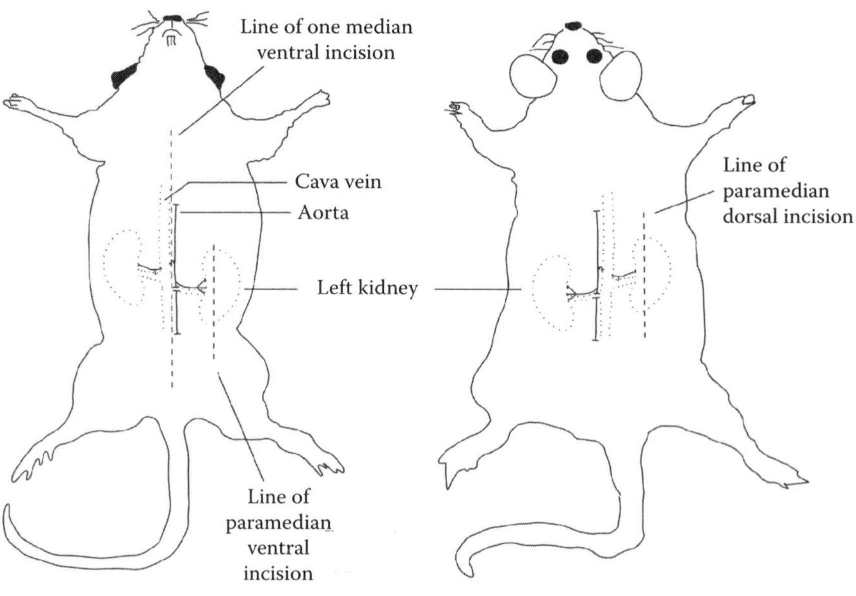

FIG. 37.1 Lines of incision for ablation of the right kidney and for the microsuture of branches of the left renal artery are illustrated.

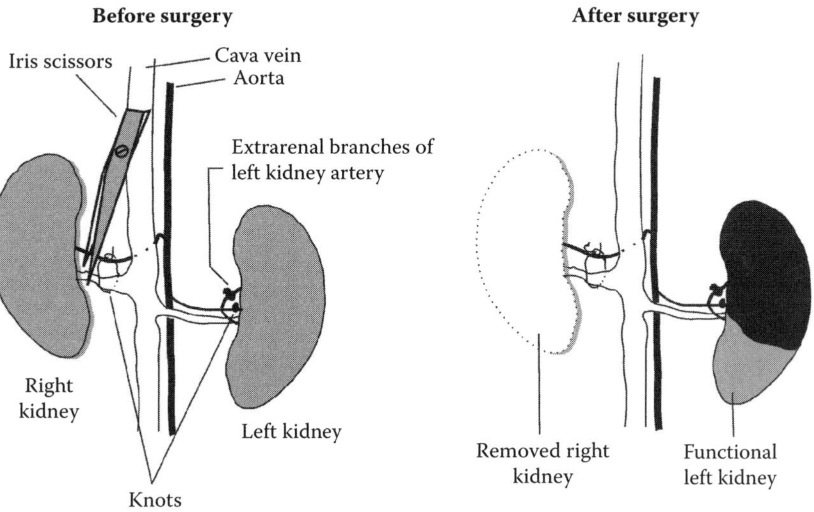

FIG. 37.2 Remnant kidney model before (left) and after surgery (right). Gray areas represent normal kidney tissue, the black area represents the ischemic kidney tissue, and the white area represents the kidney that has been removed.

be tied. Some arteries have to be released if less that one-third of the kidney has adequate blood supply.

In the case of dryness in abdominal tissue triggered by prolonged time in surgery, it is recommended to add some saline solution (1 to 2 ml).

The abdominal muscle incision is closed with a suture floss and the skin is closed using metal clips to avoid having the conscious rat remove the stitches by scratching.

The model can be adapted to produce different degrees of renal insufficiency, which can be obtained by tying different numbers of branches of the left renal artery.

After surgery, ceftriaxone is intramuscularly injected and diclofenac is subcutaneously administered.

Experimental Results

The remnant kidney model was produced in 7-week-old male Sprague Dawley rats (n = 32), where the metabolism of glucose was studied under the presence of fluoride. Sixteen animals were subjected to a simulated surgery (sham) and the other 16 to the surgery of nephrectomy (NX). In these experiments, the right kidney was subjected to ablation and only one branch of the left renal artery was tied to produce a moderate renal insufficiency. Ninety-three percent of the rats survived 60 days after surgery. Apart from parameters of glucose homeostasis (data not shown), body weight, hematocrit, 24-hour water intake, 24-hour urinary volume, urine density, blood urea nitrogen, and clearance of creatinine were used as parameters to measure renal function. Animals were housed in metabolic individual cages for 60 days and data were collected every 15 days. The changes in the values of each variable (except clearance of creatinine) throughout time were analyzed via linear regression. Comparison of the regressions between sham operated and NX rats was done through comparing the slopes and the intercepts of the lines. Clearance of creatinine was measured on the final day of the experiment. Differences were considered significant when $p < 0.05$.

Results:

- Body weight increases along the time in both groups: sham (45.71 ± 4.95 g/week) and NX (40.07 ± 3.85). Although sham operated rats grew faster than NX rats, there were no differences in slope and intercepts.

- The hematocrit did not change throughout time, but was lower in rats with nephrectomy: NX, 43 ± 1% and sham, 45 ± 1%.
- Volume of drinking water increased in both groups throughout time without differences in slope; however, the intercept was higher in nephrectomized rats than in sham operated rats: NX, 46.6 ± 6.0 and sham, 30.6 ± 4.3 ml/day.
- Although 24-hour urinary volume did not change in NX and sham operated rats throughout time, 24-hour urinary volume was higher in NX than in sham operated rats: NX, 20.1 ± 2.7 ml/day and sham, 16.0 ± 2.2 ml/day.
- Urinary density did not change throughout the time, but was lower in NX rats than in sham operated rats: NX, 1.054 ± 0.007 g/ml and sham, 1.060 ± 0.005 g/ml.
- Blood urea nitrogen was higher in NX than in sham operated rats, without changes along the time: NX, 49.4 ± 5.0 mg/dL and sham, 27.9 ± 3.4 mg/dL.
- The clearance of creatinine measured after 60 days was lower in NX rats than in sham operated rats: NX, 1.2 ± 0.3 ml/day and sham, 2.5 ± 0.9 ml/day.

REFERENCE

1. Gretz, N., and M. Strauch, eds. 1993. The remnant kidney model. In *Experimental and genetic rat models of chronic renal failure*. Basel, Switzerland: S. Karger Publishers, pp. 1–28.

Section IX

Circulatory System

38 Experimental Myocardial Infarction

Ignacio M. Seropian and Germán E. González

INTRODUCTION

The experimental model of myocardial infarction developed in rats is the most widely used throughout the world. This model closely reproduces many aspects of ischemic cardiomyopathy and heart failure, having several advantages compared with similar models performed in other species. The surgical procedure for developing myocardial infarction in the rat, as well as its uses, limitations, advantages, and disadvantages, will be described in this chapter. Surgical material is also described, plus there are tips to improve the results of the surgery. It is important to note that this surgical procedure represents a model of permanent coronary artery ligation that leads to myocardial infarction.

The ligature of one or more coronary arteries can produce permanent coronary artery occlusion with the consequent myocardial infarction. This surgical procedure was first tested in dogs in 1937. Canines were used because their coronary circulation is similar to humans and because the dog was the most convenient experimental animal for thoracic surgery at that time. However, as a test procedure, coronary occlusion in the dog produces results that are extremely variable and with limited statistical usefulness. Due to this, it was necessary to approach the problem of permanent coronary occlusion and myocardial infarction using species with more acceptable quantitative methods for reproducing this pathology. Therefore, the ligature of the coronary artery in rats and other rodents has gained high interest around the world because these species are a more standardized animal for research.

To the best of our knowledge, the first record of experimental myocardial infarction documented in rats was made in 1946 by R.F. Heimburger.[1] The advantages of using rats are that a large number of animals can be handled easily and maintenance costs are low. A rat is also a good model for reproducing more extensive and consistent infarction compared with other species. However, their chief advantages lie both in the large amount of animals that can be used in a protocol and the possibility of grouping the animals' results according to their infarct size. This allows a better understanding of the functional parameters that accompany the infarct.

UTILITY

The myocardial infarction in rats can be used for studying the histopathological temporal evolution and the evolution of functional changes along the infarct. In addition, it represents a model of ventricular remodeling and it can be used to test several drugs that modify that remodeling. Alternatively, this experimental model can be used as a cause of heart failure. However, it is also important to note that the experimental model of infarction in the rat, like other species, has important limitations to be considered. One of the limitations is that the coronary ligature represents a model of sudden artery blood flow occlusion in a previously healthy heart. This is not analogous to the pathogenesis of infarction in humans, as the model initially contained an otherwise normal myocardium and coronary arteries. Humans with myocardial infarction have developed atherosclerosis prior to the occlusion and they have many risk factors that are not present in the animal model.

Materials

- General anesthetic
- Thermostatized rat board
- Ring-handled preparation scissors
- Ring-handled iris scissors
- Cotton pliers
- Adson forceps
- Rat-tooth forceps
- Needle holder
- Microforceps
- Weitlaner–Locktite retractor (3 × 3 blunt teeth)
- Mechanical ventilator
- Endotracheal tube
- Elastic bands
- 1-ml syringe
- Suture needle ½ – 5 to 10 mm
- Suture thread
- Cotton buds
- Gauze

Procedure

Animals can be anesthetized with ketamine and xylazine intraperitoneally, using a subcutaneous needle. Then, when sedated, the chest is shaved. The rat is then placed on the rat board, lying down and stretched using elastic bands.

After tracheostomy, intubation is done with a special tube, without connecting the tube to the mechanical ventilator. It is important to note that intubation could also be performed directly from the mouth of the rat using a special tube (Abbocat 16). This tube has a curvature (approximately 20°) in the tip, allowing a better way to enter into the trachea.

For the thoracotomy, a new cut in the skin is performed in the left part of the chest. This cut should follow an imaginary straight line from the xiphoid appendix to the left armpit. Then, the muscle plane must be dissected in order to expose the thoracic wall. Subsequently, the fourth intercostal space is opened by placing a cut in the intercostal muscles. At this time, the endotracheal tube must be connected to a mechanical ventilator to avoid pulmonary collapse from pneumothorax.

In the fourth intercostal space, a chest retractor is used to open the thorax. Then, the pericardium is exposed and the pericardectomy can be performed. For this procedure, it is necessary to use an extra fine forceps.

After the pericardectomy, the heart is exteriorized and partially immobilized by its apex using an anatomical forceps. Hence, the area where the ligature will be performed is clearly exposed. The descending branch of the left coronary artery can be ligated between the left auricle and the pulmonary outflow tract using a 6.0 silk suture thread. Accordingly, the proper coronary artery ligation can be confirmed by the change in the color of the left ventricle, which turns pale. Sometimes, an increase in the heart rate is noticed.

Then, the chest is closed by putting the ribs back together with two intercostal sutures. To avoid leaving a pneumothorax, it is important to increase pressure in the subpleural space. This can be done by pressing on the thorax.

TABLE 38.1
Left Ventricular and Systemic Hemodynamic Measurements after Myocardial Infarction

	MAP	LVDP	$+dP/dt_{max}$	LVEDP	$-dP/dt_{max}$	$t_{1/2}$
			Six Hours after Surgery			
Sham (n = 7)	96 ± 7	101 ± 4	2079 ± 234	4.9 ± 0.4	1849 ± 169	8.8 ± 0.9
MI (n = 19)	80 ± 3	81 ± 3*	1670 ± 118	9.8 ± 1.3	1122 ± 76*	12.9 ± 1.0
			35 Days after Surgery			
Sham (n = 4)	95 ± 4	122 ± 3	3423 ± 175	2.6 ± 0.3	2459 ± 64	7.0 ± 0.7
MI (n = 9)	96 ± 2	103 ± 3*	2397 ± 110*	7.5 ± 1.0	1536 ± 75*	11.3 ± 1.0

* $p < 0.05$ versus sham.

Note: MAP = mean arterial pressure, LVDP = left ventricular developed pressure, LVEDP = left ventricular end diastolic pressure.

Finally, the endotracheal tube is removed and the trachea and skin are sutured. It is recommended to wait for a while before removing the tube to assure proper ventilation. Otherwise, the ventilator is replaced until the effect of the anesthesia fades. At the end of the surgery, the animals should be placed in a quiet environment to recover from the anesthesia.

Precautions

Be sure that all the pericardium is removed so there will be no adherences that may interfere with ventricular remodeling. Because the heart rate in the rat is high, special care should be taken not to tear the myocardium during passage of the needle.

Experimental Results

Myocardial infarction leads to a number of functional changes directly related to the loss of contractile mass that means the infarct size. For an infarct size of 15 to 30% (expressed as percentage of the left ventricle), a fall in both diastolic and systolic ventricular function should be observed (Table 38.1) within only 6 hours after infarction. In this case, the fall of ventricular function remains uncompensated for a long period of time, e.g., 35 days after infarction. If the infarct size is smaller (less than 15%), ventricular function can be compensated and remain normal at 35 days.

The gold standard to assess a good technique in this experimental model is to demonstrate myocardial necrosis and to produce a similar infarct size. Evaluation at 35 days after surgery can be performed by planimetry in histological samples stained with a collagen specific technique (e.g., Masson's Trichrome).[2] Six hours after surgery, infarct size can be measured by the triphenyltetrazolium chloride test (TTC).[3]

In vivo left ventricular function can be studied by using a catheterization technique (fluid-filled) where the catheter is connected to a pressure transducer and real time ventricular pressure time curves are obtained with a computer. We measured end diastolic pressure (LVEDP), $-dP/dt_{max}$, tau, $t_{1/2}$ and t_{63} as indexes of diastolic function, and $+dP/dt_{max}$ and developed pressure (LVDP) as indexes of systolic function (see Table 38.1).

REFERENCES

1. Heimburger, R.F. 1946. Injection into the pericardial sac and ligation of coronary artery of the rat. *Arch. Surg.* 52:677–689.

2. González, G.E., J. Palleiro, S. Monroy, S. Pérez, M. Rodríguez, A. Masucci, R.J. Gelpi, and C. Morales. 2005. Effect of the early administration of losartan on the functional and morphological aspects of post-myocardial infarction ventricular remodeling in rabbits. *Cardiovasc. Pathol.* 14(2): 88–95.
3. Donato, M., V. D'Annunzio, G. Berg, G.E. González, L. Schreier, C. Morales, R.L. Wikinski, and R.J. Gelpi. 2007. Ischemic postconditioning reduces infarct size by activation of A1 receptors and K^+ (ATP) channels in both normal and hypercholesterolemic rabbits. *J. Cardiovasc. Pharmacol.* 49(5): 287–292.

39 Experimental Arteriosclerosis

Anabel Brandoni and Adriana M. Torres

INTRODUCTION

The current concept of arteriosclerosis is primarily influenced by experimental observations of rabbits maintained on an atherogenic fatty acid/cholesterol-rich dietary regime. Nevertheless, the rat has proved to be useful for pathophysiological studies on different types of arteriosclerosis for which no satisfactory animal model had hitherto existed.[1]

MODEL OF ARTERIOSCLEROSIS BY CALCIUM OVERLOAD PRODUCED WITH VITAMIN D_3

There is evidence that vascular calcium overload is an important pathogenic factor in many vascular diseases, such as hypertension, atherosclerosis, and arteriosclerosis. This pattern of rat arteriosclerosis corresponds to the Mönckeberg type of media calcinosis,observed in human pathology. Old and/or hypertensive rats have been used to study arterial calcinosis, but these are not ideal models because, in animals, vascular calcium overload linked with age or hypertension is less pronounced than in humans. The greatest degree of Ca^{2+} overload with simultaneous structural damage of the arterial walls was achieved by administration of high doses of vitamin D_3 or dihydrotachysterol in rats. Under these conditions, Ca^{2+} overload occurred within a few days.

In our laboratory, an increase in the calcium content of the arterial wall was produced in adult rats following treatment with vitamin D_3. In this experimental model, calcium overload is not restricted to the arterial wall, and it has been demonstrated in many organs, such as heart and kidneys, leading to the impairment of their function.[2-4]

MATERIALS

- General anesthetic
- Thermostatized rat board
- Ring-handled preparation scissors
- Ring-handled iris scissors
- Cotton pliers (2)
- Adson forceps
- Artery forceps
- Rat-tooth forceps
- Needle holder
- Spatula
- Catheter
- 1-ml syringe
- Injection needle 15 × 0.5
- Saline solution
- Heparin
- Iodopovidone

Suture needle ½ – 5 to 10 mm
Suture thread
Cotton buds

Procedure

Rats are injected with vitamin D_3 (300,000 IU/kg body weight, intramuscular (i.m.)). Vitamin D_3 solutions are prepared in corn oil. Control animals receive an identical volume of corn oil via i.m. injection. Animals are allowed free access to a standard diet and tap water.

Experimental Results

Adult male Wistar rats were randomly divided into four groups: control (Group C) and vitamin D_3-treated rats 5 days (Group T5), 10 days (Group T10), or 20 days (Group T20) before the experiments.

The day of the experiment, systolic arterial pressure was measured using a Harvard indirect rat-tail blood pressure monitor (Harvard Apparatus, Millis, MA, U.S.A.) connected to a Harvard student oscillograph.

The animals then were anesthetized with sodium thiopental (70 mg/kg body weight, intraperitoneal (i.p.)) and placed in a thermostatic stretcher so as to maintain a constant body temperature. Samples of the abdominal aorta were removed for the analysis of tissue calcium levels. Tissues were weighed and then heated to constant dry weight for 48 hours at 120°C. Dry tissue samples were dissolved in nitric acid (14 M) and left for 72 hours at room temperature. The samples were then centrifuged (2,000 × g, 10 minutes) and the supernatant removed. Strontium nitrate was added and calcium (μmol/g dry weight) was measured by atomic absorption spectrophotometry.

Systolic arterial pressure and total calcium levels in aortic tissues were increased in all treated groups, as shown in Table 39.1.

TABLE 39.1
Systolic Arterial Pressure and Total Calcium Levels in Aortic Tissues in Control and Vitamin D_3 Treated Rats 5 Days (Group T5), 10 Days (Group T10), or 20 Days (Group T20) Before the Experiments

	Systolic Arterial Pressure (mm Hg)	Total Calcium in Aorta (μmol/g Dry Weight)
Control	96 ± 3 (n = 9)	23 ± 1 (n = 9)
T5	158 ± 7[a] (n = 12)	36 ± 6[a,c,d] (n = 4)
T10	136 ± 8[a] (n = 8)	50 ± 5[a,b] (n = 7)
T20	141 ± 13[a] (n = 7)	55 ± 3[a,b] (n = 6)

[a] $p < 0.05$ versus control.
[b] $p < 0.05$ versus T5.
[c] $p < 0.05$ versus T10.
[d] $p < 0.05$ versus T20.

Note: The results are expressed as mean ± SEM. Statistical analysis were performed using analysis of variance and Newman–Keuls posttest.

OTHER MODELS OF ARTERIOSCLEROSIS

1. *Arteriosclerosis of spontaneously hypertensive Okamoto rats (SHRS)*: This model closely resembles the vascular damage associated with essential hypertension in humans. SHRs show marked proliferation of media cells and migration into the subendothelial layers accompanied by diffuse calcification on mesenteric artery branches.
2. *Arteriosclerosis of hypertensive salt-sensitive DAHL rats under an 8 wt% NaCl-containing diet*: This model is characterized by early development of Ca^{2+} and lipid-containing intimal plaques with endothelial desquamation, and the lining of the luminal side of the arterial walls with a layer of fibrin.
3. Models of human risk factors: The natural aging process of human arteries is reflected in an augmentation of Ca^{2+} content. Smokers and diabetics are more prone to arteriosclerosis and arterial calcinosis than the normal population. There is evidence that severe diabetics and heavy smokers exhibit a dramatic acceleration of the natural age-dependent process of spontaneous arterial calcium accumulation so that, for example, in uncontrolled severe diabetics, the development of arterial calcinosis and its pathological consequences may be anticipated by as much as 25 years.

These human risk factors can also be modeled in rats. Alloxan-diabetic rats and rats fed chronically with nicotine (particularly when combined with dihydrotachysterol or vitamin D_3) accumulate arterial Ca^{2+}. Rats fed a fatty acid/cholesterol-rich diet for several months also develop arterial Ca^{2+} overload; this provides a useful model for problems associated with the Western diet.

Experiments using these rat models indicate that arterial Ca^{2+} overload is fundamentally involved in the pathogenesis of arteriosclerosis precipitated by a variety of different causes, and probably constitutes a common etiological denominator.[1]

REFERENCES

1. Fleckenstein, A., M. Frey, J. Zorn, and G. Fleckenstein-Grün. 1987. The role of calcium in the pathogenesis of experimental arteriosclerosis. *Trends Pharmacol Sci.* 8: 496–501.
2. Quaglia, N.B., C.G. Hofer, and A.M. Torres. 2002. Pharmacokinetics of organic anions in rats with arterial calcinosis. *Clin. Exp. Pharmacol. Physiol.* 29: 48–52.
3. Quaglia, N.B., A. Brandoni, A. Ferri, and A.M. Torres. 2003. Early manifestation of nephropathy in rats with arterial calcinosis. *Ren. Fail.* 25: 355–366.
4. Quaglia, N.B., A. Brandoni, S.R. Villar, and A.M. Torres. 2004. Haemodynamic and tubular renal dysfunction in rats with sustained arterial calcinosis. *Clin. Exp. Pharmacol. Physiol.* 31: 231–236.

40 Isolated Heart
Langendorff Technique

Bruno Buchholz, Veronica D'Annunzio, and Ricardo J. Gelpi

INTRODUCTION

The history of the isolated perfused heart starts in the nineteenth century when Carl Friedrich Wilhelm Ludwig designed a method for isolated mammalian heart artificial nutrition. In that model, the aorta was tied into the carotid artery of a conscious animal. The blood of the donor animal perfused the coronary arteries of the receptor heart. However, this method gained no major popularity, probable due to coagulation problems.

Twenty years later, together with Elias Cyon, the two developed the first perfusion of isolated heart from a frog in order to assess the effects of the temperature on cardiac activity. In 1883, Henry Newell Martin developed an isolated mammalian heart perfuse system (using dogs and cats) with intact lung circulation, mechanically ventilated lungs, and artificial peripheral circulation. In this model, the preload and the afterload were variable. In 1895, Oscar Langendorff finally succeeded in isolating and perfusing the mammalian heart, which continued to beat and pump blood for several hours. Numerous researchers have improved and modified Langendorff's experimental model in the subsequent decades. Among the improvements, only the assessment of cardiac behavior force using the measurement of left ventricle function will be emphasized.

MATERIALS

General anesthetic
Thermostatized rat board
Ring-handled preparation scissors
Ring-handled iris scissors
Adson Forceps
Artery forceps (2)
Cannula for Langendorff
Latex balloon
1-ml syringe
Suture needle ½ – 5 to 10 mm
Suture threads
Heparin
Register and acquisition data equipment

PROCEDURE

The animal is anesthetized with a mixture of ketamine and xylazine (see Chapter 4). Once achieving the anesthetic effect, the animal is placed in dorsal decubitus on the rat board. A longitudinal medial incision is made in the ventral face of the thorax. Both sides of the skin are gripped with artery forceps. By divulsion of the subjacent tissue with an iris scissors, the pectoral muscle is exposed. These muscles are sectioned cross-sectionally to expose the thorax. Following that, a lateral cut is performed on the chest on both sides to raise the entire anterior thorax. Thus, the intrathoracic organs are exposed. A fast divulsion of the elements that surround the supracardiac vessels is performed, and then they are sectioned with scissors through a neat cut passing through the least upper bound of the ascending portion of the aortic arch. The heart is isolated by sectioning the tissue adherences, extracted from the animal, and placed into an iced box. An assistant is required for this procedure. The assistant takes the end of the aorta with two small artery forceps. Then, the surgeon introduces the cannula and fixes the aorta to the end of the cannula with suture thread. Because there is a possibility that the cannula may cause a valvular insufficiency, it is important to mention that the cannula must not be introduced into the ventricular chamber. During the cannulation process, the heart is perfused in a retrograde way with a mixture of heparin (20 IU/ml) and Krebs–Henseleit solution for a total volume of 1 ml. This procedure is performed to wash the blood out of the coronary tree and to avoid the formation of clots. All of these procedures must be performed in 60 seconds or less to avoid heart deterioration.

Once extracted, the heart is placed, through the cannula, in an isolated perfusion system according to the modified Langendorff technique (Figure 40.1).

The heart is perfused with Krebs–Henseleit solution, which is composed of NaCl 118 mM, $CaCl_2$ 2 mM, KCl 5.9 mM, $MgSO_4$ 1.2 mM, $NaHCO_3$ 20 mM, and dextrose 11.1 mM. The solution is balanced with a mixture of 95% O_2 to 5% CO_2 in order to oxygenate and to maintain a pH 7.4, at 37°C.

Once the heart is placed on the isolation perfusion system, it is necessary to cut the left atria to be able to introduce a latex balloon that is connected to a polyethylene tube into the left ventricular chamber, crossing the mitral ring. The distal extreme of the tube is connected to a pressure transducer, which allows for the measurement of the intraventricular pressure (Figure 40.2).

Another pressure transducer is placed before the aortic cannula to measure the coronary perfusion pressure (CPP).

The left ventricular pressure and CPP are registered online using register and acquisition data equipment. The coronary flow is controlled by a peristaltic pump, and must be increased in a gradual way to obtain a CPP of approximately 70 mmHg. The CPP is maintained constant during the entire experiment.

In the right atrial, two electrodes connected to an external pacemarker are sutured to the left and right atrial to maintain the constant heart rate during the experiment, in a value of 300 beats per

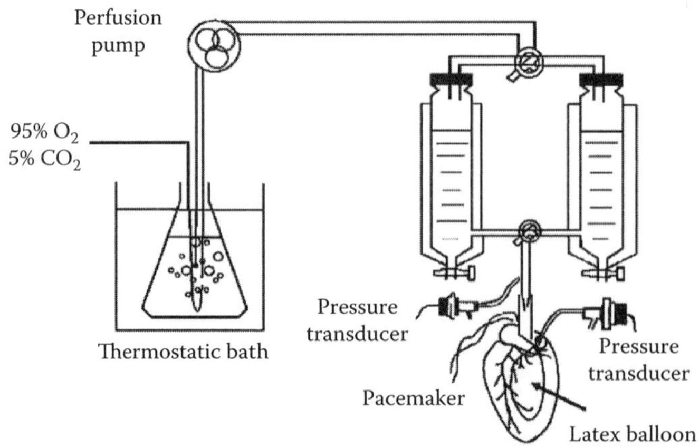

FIGURE 40.1 Schematic graph of the perfusion system is illustrated, according to the Langendorff technique, which is in use for the accomplishment of the experiments.

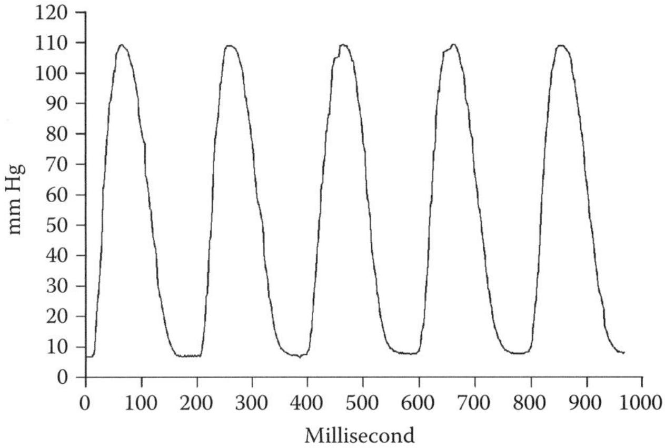

FIGURE 40.2 Intraventricular pressure is shown.

minute. Then, the system (or experimental model) is stabilized for 20 minutes before beginning the experimental protocol; in this way, it is possible to keep it stable for approximately 2 hours.

Advantages and Disadvantages

The isolated heart provides a highly reproducible experimental model, which can be studied quickly and in large numbers of species. It allows a broad spectrum of physiological and pharmacological indices to be measured. These measurements can be made in the absence of the confounding effects of other organs, the systemic circulation, and a host of peripheral complications, such as circulating neurohormonal factors. While the fact that the isolated heart is denervated must always be taken into account, this can be turned into an advantage, allowing the separation of cardiac from sympathetic and vagal stimulation. Denervation and the absence of other peripheral factors can often be compensated for, thus catecholamines or other neurotransmitters may be included in perfusates and many other peripheral factors can be added exogenously and in a controlled manner, which again can represent a significant investigational strength. Certainly, the isolated perfused heart provides an excellent test bed for undertaking carefully controlled dose-response studies of metabolic or pharmacological interventions.

It must be recognized that, as an *ex vivo* experimental model (or experiment), the isolated heart is a constantly deteriorating system (or experimental model), but, nonetheless, it is capable of being studied for several hours. The model also allows the induction of the whole heart or regional ischemia and this can be achieved at various levels of flow. Similarly, anoxia or hypoxia at various degrees of oxygen deprivation (in the presence of normal flow) can be easily imposed. The isolated heart model is amenable to reperfusion or reoxygenation at various rates and with various reperfusate compositions, thus providing a powerful tool for assessing many aspects of ischemia- and reperfusion-induced injury. Arrhythmias are readily induced and studied, especially in the larger hearts where conduction pathways can be mapped and a variety of electrophysiological recordings can be made. It is important to consider that the isolated heart model allows experiments to be continued in the face of events (e.g., infarction-induced loss of pump function, cardiac arrest, or arrhythmias), which would normally jeopardize the survival of an *in vivo* experiment.

Experimental Results

Figure 40.3 shows the behavior of left ventricular function in an isolated rat heart perfused according to the Langendorff technique during 30 minutes. Left ventricular developed pressure (LVDP)

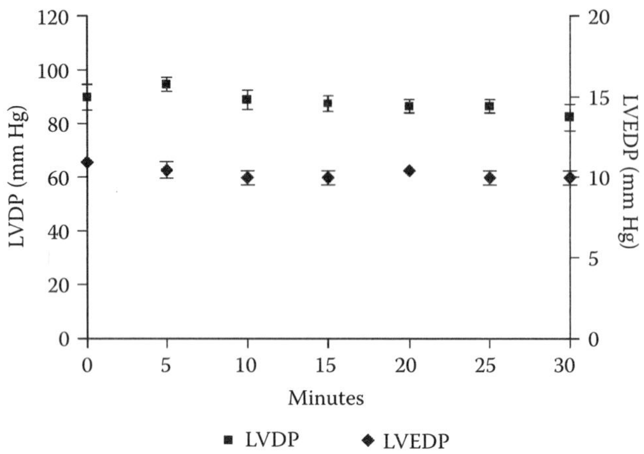

FIGURE 40.3 Left ventricular developed pressure (LVDP) and left ventricular end diastolic pressure (LVEDP) are represented in left and right panels, respectively. The results are expressed as mean ± SEM.

decreased only 8.57% compared with basal values. Furthermore, no changes in left ventricular end diastolic pressure (LVEDP) were observed after 30 minutes of perfusion.

Table 40.1 shows LVDP and LVEDP values at baseline and during different reperfusion times. The isolated rat hearts were perfused according to the Langendorff technique. In all groups, 30 minutes of global ischemia and 2 hours of reperfusion were performed. In the postconditioning group, after 30 minutes of global no-flow ischemia, reperfusion was started for 30 seconds followed by 30 seconds of ischemia, repeated for two cycles (2 minute total intervention, ischemic postconditioning protocol). LVDP was significantly lower compared with the respective preischemic values, but showed no significant group-related differences throughout the procedure. LVEDP (myocardial stiffness) increased during the reperfusion period in all groups, but there were no significant differences between groups.

Figure 40.4 represents the infarct size after 30 minutes of global no-flow ischemia, expressed as a percentage of the total left ventricular area. In control hearts, the infarct size was 16.6 ± 1.5%. Ischemic postconditioning significantly reduced the infarct size to 5.4 ± 1.3%.

In conclusion, ischemic postconditioning significantly reduces infarct size without changes in left ventricular function during reperfusion.

TABLE 40.1
Ventricular Function; Left Ventricular Developed Pressure and Left Ventricular End Diastolic Pressure are Represented.

		Basal	10 Minute Reperfusion	20 Minute Reperfusion	30 Minute Reperfusion
LVDP	Control	92.02 ± 3.15	28.18 ± 3.37*	34.53 ± 3.45*	38.45 ± 3.16*
(mmHg)	Postcon	98.81 ± 17.01	31.37 ± 3.11*	40.09 ± 3.73*	42.13 ± 3.85*
LVEDP	Control	10.22 ± 1.14	53.42 ± 4.58*	50.86 ± 4.56*	48.85 ± 4.68*
(mmHg)	Postcon	10.62 ± 1.55	60.01 ± 5.29*	58.21 ± 5.30*	57.39 ± 5.73*

* p <0.05 versus, respectively basal value.

Note: The results are expressed as mean ± SEM. Postcon: ischemic postconditioning.

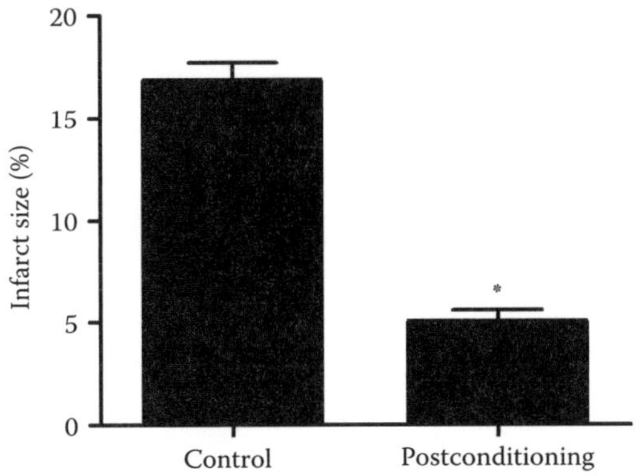

FIGURE 40.4 Infarct size is expressed as a percentage of left ventricle area. The results are expressed as mean ± SEM. *= $p < 0.05$ versus control, control: 30 minutes of global no-flow ischemia.

Section X

Additional Techniques

41 Polyclonal Antibodies Preparation

Laura I. Pera, Lucas R. M. Brun, and Alfredo Rigalli

INTRODUCTION

Polyclonal antibodies are produced by different clones of B cells in a phylogenetically different animal from that where the antigen is derived. Monoclonal antibodies, on the other hand, are produced *in vitro* by a unique clone of isolated B cells. Although monoclonal antibodies have some advantages over polyclonal antibodies, the latter have higher affinity and reactivity against different epitopes of the same antigen.

For the production of polyclonal antibodies, immunization it is needed, which entails the injection of an antigen. The antigen must be injected in adequate amounts to obtain a proper response and the purity of the antigen is also important. Different techniques are available for purification of antigens, such as precipitation, gel filtration, ultrafiltration, affinity chromatography, electrophoresis, ionic chromatography, etc. The combination of two or more of these is often required to obtain a suitable purity.

The amount of antigen to be injected depends on the animal where the antibody will be raised. For mice, 0.05 to 1 mg is adequate and up to 5 mg in the case of a goat. In the rat and guinea pig, the amount of antigen ranges from 0.05 to 0.2 mg. Small antigens have little effect on the immune system, so that they must be administered together with adjuvants or joined to other molecules to induce an adequate immune response.

UTILITY

This technique is useful for the production of polyclonal antibodies. They can be used for the recognition of proteins in Western blotting or immunohistochemistry, for example.

Materials

General anesthetic
Purified protein
Complete Freund's adjuvant
Incomplete Freund's adjuvant
1-ml syringe
Injection needle 25 × 0.8
Eppendorf tubes
Centrifuge tubes
Centrifuge

Procedure

One hundred µl of solution containing 0.05 to 0.2 mg of purified antigen is mixed with 0.5 ml of complete Freund's adjuvant (CFA). Both liquids are immiscible and an emulsion must be completed before injecting. The emulsion is performed with the aid of a syringe and a 25 × 0.8 injection needle. The mixture is forced to flow in and out of the syringe repeatedly until the mixture has a white color. Five to six times is enough to obtain a stable emulsion. The same procedure must be done to obtain the emulsion using the incomplete Freund's adjuvant (IFA) for the following immunizations or "boosters." The number of boosters is evaluated based on the immune response and the concentration of antibodies found in the serum (see Experimental Results below).

Precautions

Wearing protective glasses and gloves is important when working with the emulsion because the material is very hazardous, especially for the eyes. Eye contact with CFA can produce severe inflammation.

Detection of Antibodies by Immunodiffusion

Precipitation in agar–agar is a technique of moderate sensitivity in comparison with enzyme-linked immunosorbent assay (ELISA) or radioimmunoassay (RIA). However, it is widely used because of its versatility, low cost, and easiness. Ouchterlony's technique[1] detects the presence of antibodies by precipitation of the antigen–antibody complex. The reaction is made in a thin layer of agar–agar using a microscope slide or a Petri dish as a holder. Antigen solutions and sera are placed in holes done in the agar at specific distances from each other. Solutions diffuse in a radial way and a concentration gradient of antigen and antibody is established. Between the holes there is an equivalence of antigen and antibody where the complex precipitates, as a curved line (Figure 41.1).

The shape and location of the precipitation line is determined by the concentration and molecular mass of antigens and antibodies. The precipitation line is closer to the lower concentration reactant and concave to the higher molecular mass molecule. Although precipitation of the antigen–antibody complex occurs in the equivalence region, an inadequate concentration ratio will avoid precipitation and visualization.

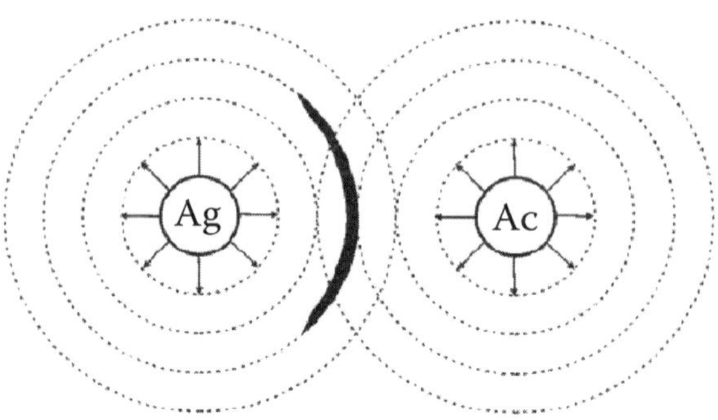

FIGURE 41.1 Antigen–antibody reaction is illustrated.

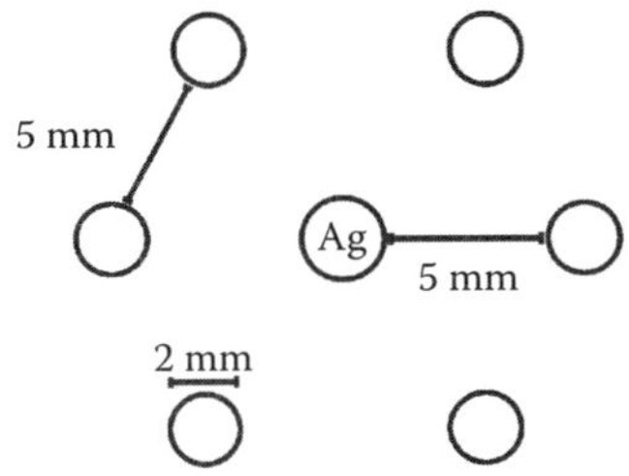

FIGURE 41.2 Separation between holes in the agar–agar is shown.

Procedure

Agar–agar 1.5% w/v in phosphate buffered saline (PBS) is used for making the thin layer on a slide for microscopy. Agar–agar must be dissolved in PBS at 80 to 90°C. The process takes as long as 1 hour, and at the end, the solution will be transparent. The slide is placed on a horizontal surface to obtain a regular thickness of the agar–agar layer. Irregularities in the thickness of the layer will cause modification in the diffusion rate of antigen and antibody. If the holder is a 2 × 6-cm-microscope slide, 2 ml of agar–agar solution is adequate. This volume is placed with a pipette on the slide, which has been previously cleaned with ethylic alcohol. The same pipette is used to uniformly distribute the agar–agar on the slide until a 2-mm-thick layer is formed. The process must be done quickly before the agar–agar solidifies, which can produce a layer of different thickness. The slide with agar–agar is stored in a container with high humidity, such as a Petri dish with filter paper, soaked with water.

The holes for antigen and antibody must be located 5 mm apart and have a diameter of 2 mm. A scheme with the holes must be created on paper. Then the slide with agar–agar is superimposed on the scheme, and the holes are made with a 2 mm tube adapted to a vacuum pump. When the antigen is to react with different samples containing the antibody, it is convenient to make a central hole for the antigen, surrounded by holes where the samples or their dilutions are placed. Five μl of the solution of purified antigen are pipetted in the central hole. Five μl of each sample is pipetted in the surrounding holes, located at 5 mm from the central hole (Figure 41.2).

Different dilutions of serum are often needed for an adequate visualization of the precipitation lines. Once the slide is placed in the Petri dish, it is maintained for 48 hours at 4°C. After that, if the serum has the appropriate antigen concentration, a line of precipitation will appear between the holes. Once the lines of precipitation are detected, the agar–agar can be washed with PBS to eliminate the nonprecipitated proteins. The slide is covered with PBS for 1 hour with constant shaking. The process is repeated four times. Finally, the agar–agar is dyed with Coomassie brillant blue, photographed, and digitalized.

Experimental Results

Immunization to obtain antirat α-macroglobulin (anti-αM) antibody was performed in two male adult guinea pigs. Blood was withdrawn by cardiac puncture before immunization and after each booster. Samples were mixed with heparin and, after separation of blood cells, plasmas were frozen at –20°C.

An emulsion of 100 μl of solution containing 0.06 mg of rat α-macroglobulin and 0.5 ml of CFA was obtained.

Inoculation scheme: Previous to inoculation, a sample of blood was obtained (sample M1). After that, the injection of the emulsion was subcutaneously and intramuscularly performed in different places of the body.[2] After 15 days, a sample of blood was obtained (sample M2) and saved for checking the presence of anti-αM antibodies. Guinea pigs were boosted after 30 days with an emulsion of 100 μl of αM and 1 ml of IFA. After 20 days, a sample of blood was obtained (sample M3). Two more boosters were performed as stated above, and samples of blood were collected (samples M4 and M5).

In order to verify the immune response of the guinea pig, Ouchterlony's technique was employed.[1] Five μl of purified αM were placed in the central hole of the agar–agar. In the surrounding holes, 5 μl of samples were poured (M1, M2, M3, M4, and M5).

REFERENCES

1. Ouchterlony, O. 1948. Antigen–antibody reactions in gel. *Acta Pathol. Microbiol. Scand.* 26: 507–511.
2. Margni, R.A. 1977. Preparación de sueros inmunes y purificación de anticuerpos. In *Inmunología e inmunoquímica. Fundamentos*. Segunda edición. Buenos Aires, Argentina: Editorial Panamericana. Bs.As.

42 Histological Procedures

María L. Brance and Lucas R. M. Brun

INTRODUCTION

The histological techniques are useful and can help in the surgery with animals. They verify that the ablation of an organ has been done successfully or that a treatment has produced the desired effect on a particular tissue. A brief description of the basic histological techniques will appear in this chapter.

The examination of the structures of cells, tissues, or organs can be made immediately with or without dyes. Cell or tissue structures can also be observed microscopically after fixing and dying.[1,2]

Immediate examination is done microscopically over living elements of parts that survive for a definite time. It can be done without any substance that modifies its aspect or with a vital stain. The latter can be classified as intravital staining, when the dye is introduced in the animal while alive, or supravital staining, when the dye is added to structures that are preserved alive for a definite time, but outside the body of the animal. These kinds of examinations are done especially when observing protozoa, erythrocytes, leucocytes, and spermatozoids. The specimen is put between a slide and a glass cover with a drop of the liquid where the specimen comes from or another fluid, such as plasma, serum, or isotonic solutions. In order to avoid dehydration, the glass cover must be sealed with petroleum jelly or paraffin. To obtain long-life slides, it is necessary to apply definite techniques, which are described below in a practical way.

OBTAINING THE SPECIMEN

1. A small piece of tissue is obtained with the aid of a scalpel. Special care should be taken not to damage the sample.
2. The specimen is cut in pieces no thicker than 5 to 10 mm.
3. Small pieces of the specimen are put in special plastic cages.
4. Cages with the specimen are put in containers with fixative liquid in order to avoid the autolysis of the tissue that occurs in postmortem state. The fixative can be a physical or a chemical treatment. Wet and dry heat, dryness and freezing are examples of physical fixatives. Chemical treatments can be divided into simple and complex. Formaldehyde, osmium tetroxide, ethylic alcohol, and acetic acid are examples of simple chemical treatment. Bouin and Zenker fixatives are examples of the latter. The most common fixative is a solution of 10% w/v buffered formaldehyde (pH 7), and the specimen must be immersed in it for 24 to 72 hours. The ratio volume formaldehyde/volume specimen is approximately equal to 50. The immersion in formaldehyde disrupts the cellular metabolism, and preserves the cellular and extracellular structures for further processing of samples.[3]
5. After fixative action, profuse cleaning with water is done to eliminate any remaining fixative. A small amount of formaldehyde could alter the process of staining.
6. Finally, the specimens can be stored in 70% ethylic alcohol. However, the best way to preserve the specimen is to embed it in paraffin.

Paraffin Embedding

In this step, paraffin goes into the tissues giving them elasticity and resistance, both properties needed for cutting. After the treatment with fixative, specimens are hydrated, and as paraffin is a hydrophobic substance, dehydration of the specimen and impregnation with a solvent of paraffin should be done. This process comprises the following steps.

Dehydration

This process involves the treatment with ethylic alcohol and finally acetone. The time the tissue should be in each solution will depend on the size of the specimen and on the fixative that was used in the previous steps. On average, 30 minutes for soft tissues and 60 minutes for hard tissues are recommended (Figure 42.1).

Clarification

Toluene or xylene (also known as xylol) are used as solvents of paraffin, and make it easier to impregnate the tissue. On average, 30 minutes for soft tissues and 60 minutes for hard tissues are recommended (Figure 42.2).

Embedding

Cages with the tissue specimens are placed in liquid paraffin at 60°C for approximately 15 hours. After that, the cages are put in other containers with paraffin for 3 to 4 hours, at the same temperature. In order to avoid delay in the process, it is convenient to melt the paraffin the day before the embedding step (Figure 42.3).

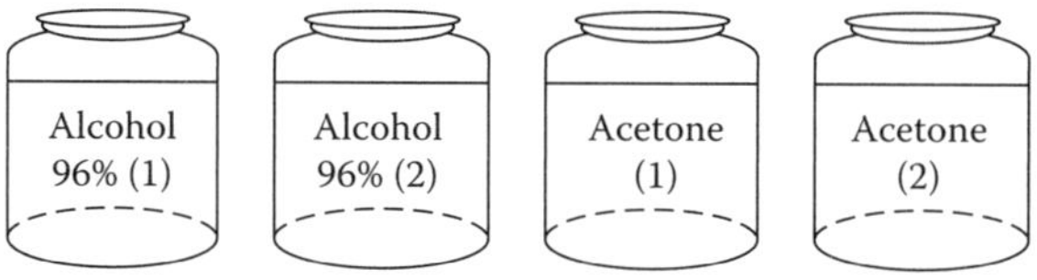

FIGURE 42.1 From left to right, numbers (1) and (2) indicate the order in which substances must be used any time the process of dehydration is required. In this way, contamination of liquids is avoided.

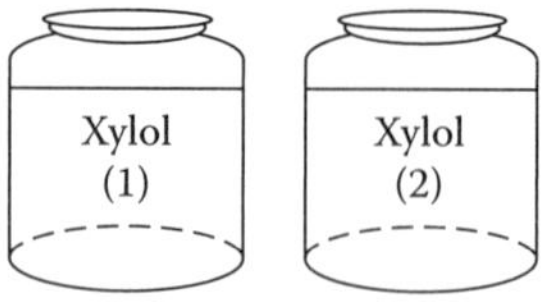

FIGURE 42.2 From left to right, numbers (1) and (2) indicate the order in which substances must be used any time the process of clarification is required. In this way, contamination of liquids is avoided.

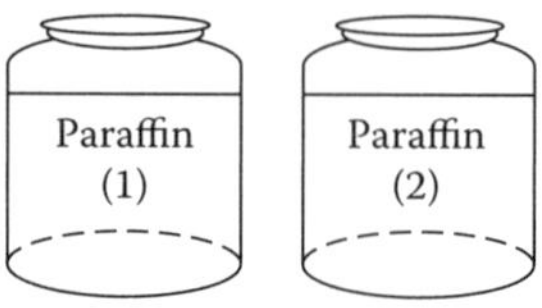

FIGURE 42.3 From left to right, numbers (1) and (2) indicate the order in which substances must be used any time the process of embedding is required.

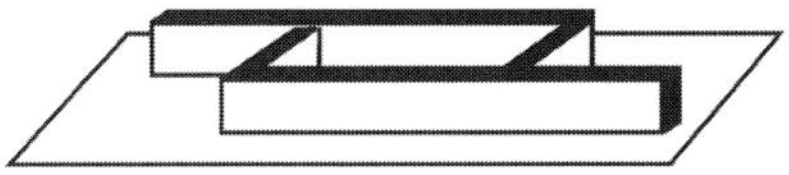

FIGURE 42.4 Leuckart's bars for paraffin block preparation are shown.

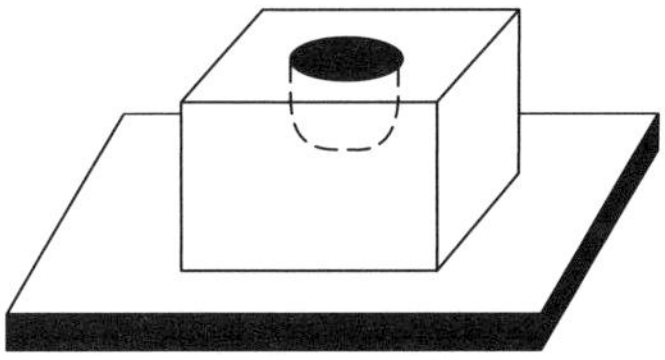

FIGURE 42.5 Illustrated is sticking the paraffin block to the wood base.

Paraffin Block Preparation

In this step, a solid block containing the tissue specimen is obtained. The aim of this step is to fix the specimen in a definite position inside a firm medium. This process maintains the tissue specimen intact, while cutting in very thin sections. The following instructions are to be followed:

1. Liquid paraffin at 60°C is poured in a box made of paper or metal until the box is full of paraffin. Leukart's bars are useful as molds for obtaining the paraffin block. They consist of two L-shaped metal bars and a flat metal surface (Figure 42.4).
2. Tissue specimens are put at a distance of 3 cm from each other, at the bottom of the block. This process is done while the paraffin is still liquid.
3. The block is allowed to lose heat until the paraffin solidifies. The time needed for this process is dependent on the environmental temperature. As a reference, 15 minutes are required in winter and about 30 minutes in summer.
4. After the paraffin has solidified, it is separated from the mold and cut with a knife that has to be heated in the flame of a Bunsen burner. The block is cut so as to avoid touching the tissue. A convenient size is 1 cm long and 0.5 cm wide.
5. The side of the block, opposite to the tissues is melted with the heated knife and, with the aid of melted paraffin, the block is joined to a base made of wood. The latter must have an appropriate size because it is the piece that joins the paraffin block to the microtome (Figure 42.5).

Cut

Once the tissue specimen has been embedded in paraffin and mounted in an adequate holder, it is cut with the microtome. The block previously cooled at 4°C is cut with the microtome until the excess of paraffin is eliminated. This procedure can be done every 15 to 20 μm. Once the tissue is on the surface of the block, 4 to 6 μm is selected to obtain the paraffin ribbon.

The slices are put in a container with water at 40°C. The warm water helps to eliminate a coarse surface that can prevent the visualization of the structures.

Adhesion

The tissue specimen slides are stuck to the slide with Mayer´s glycerinated albumin. Slides must be maintained on their sides over absorbent paper for 15 to 20 minutes and then heated at 40 to 45°C in an oven, to adhere the specimen to the slide. After that, slides are ready for staining.

FIGURE 42.6 From left to right, numbers (1) and (2) indicate the order in which substances must be used any time the process of rehydration is required. In this way, contamination of liquids is avoided.

Staining

Before staining, paraffin must be eliminated because it prevents dye from reaching the tissue. Paraffin adhered to the histological slide is eliminated by subsequent treatments with xylol, acetone, and ethylic alcohol (Figure 42.6). Slides are put in xylol (1) for 10 minutes, and then passed through the other mentioned liquids for a few seconds.

Once the paraffin is removed, slides are stained. It is important to take into account that some additional steps can be added for some special staining.

On the basis of the objective of the study that is being carried out, different dyes can be used, e.g., common staining, such as hematoxylin-eosin, or special, such as histochemistry or immunohistochemistry.

The following techniques are described below:

1. Hematoxylin-eosin staining
2. Masson's trichrome technique
3. Histochemical assay for the determination of intestinal alkaline phosphatase
4. Immunohistochemistry for the determination of intestinal alkaline phosphatase

Mounting

A cover glass is adhered to the slide with Canada balsam in order to preserve the colored tissue and to avoid refraction alterations.

Staining Techniques

Hematoxylin Eosin Technique

As mentioned above, paraffin needs to be eliminated by successive treatments with xylol, acetone, and ethylic alcohol. After this step, the specimens are completely dehydrated and they need to rehydrate by successive treatments with acetone and ethylic alcohol.

After immersion in water for 5 minutes, staining with hematoxylin is performed. This stain dyes the cellular nucleus a violet color. Because hematoxylin is originally red, a change to violet can be done by neutralizing the acidity with water for approximately 5 minutes.

Staining with 1% w/v hydrosoluble eosin is then performed. This is an acid dye, so it stains the cytoplasm of cells and intercellular substances in a red-orange color. The action of eosin is stopped by treatment with water. The time needed for the treatment with water depends on when the dye was prepared. While 1 to 5 minutes is adequate for hematoxylin, for eosin, a few seconds up to 1 minute is long enough.

Once slices have been stained, they are again subjected to dehydration by successive treatments with ethylic alcohol and acetone. After that, the immersion of the slide in buthyl acetate before the clarification with xylol is recommended. Finally, mounting is performed, as described above (Figure 42.7).

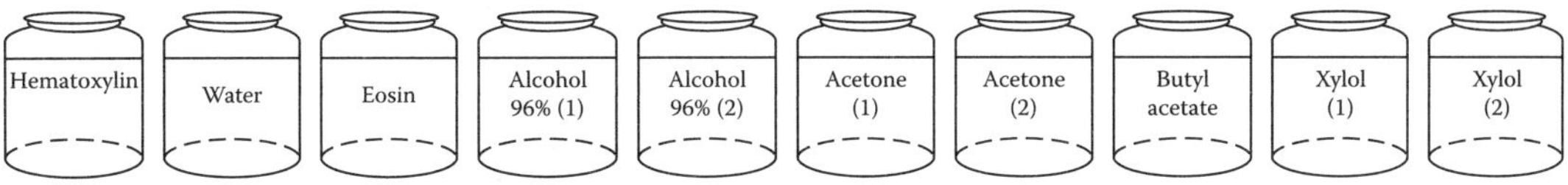

FIGURE 42.7 From left to right, solutions for hematoxylin eosin staining. Numbers (1) and (2) indicate the order in which substances must be used any time the process is required. In this way, contamination of liquids is avoided.

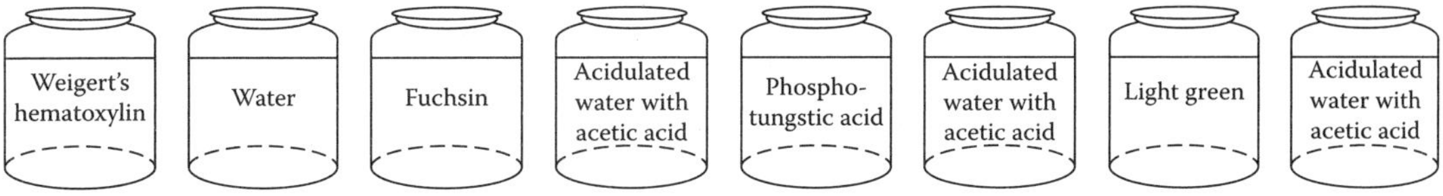

FIGURE 42.8 From left to right, solution for Masson's trichrome technique.

Masson's Trichrome Technique (Modified by Goldner–Foot)

For this technique, the following dyes are used: Weigert's hematoxylin, ponceau-fuchsin, and light green. As a result, the nucleus is dyed a black color, the cytoplasm a reddish, the collagen and the mucine green, and the erythrocytes yellow or orange.

The procedure is the same as hematoxylin-eosin. After the elimination of paraffin, the slices are stained with Weigert's hematoxylin for 3 to 5 minutes and finally washed with tap water. Immediately, the slices are immersed in the fuchsin solution for 5 minutes and then they are washed with 0.2% w/v acetic acid. After that, treatment with 5% w/v phosphotungstic acid is performed and treatment with diluted acetic acid for 5 minutes is required for developing color. Finally, dehydration, clarification, and mounting are performed as described above (Figure 42.8).

Histochemical Determination of Intestinal Alkaline Phosphatase Activity

Histochemical techniques[4,5] are useful to detect enzymatic activity. Preparation of the block, cut, and adhesion is the same as the technique for hematoxylin-eosin staining. After that, deparaffination and hydration for 5 minutes is required. Then, the slides are immersed for 30 minutes in a solution of 5-bromo-4-chloro-3-indolyl phosphate (BCIP).[6] BCIP is a substrate for alkaline phosphatase, and once hydrolizated by the enzyme, it produces an indigo insoluble compound. Finally, the excess of BCIP is eliminated with water. Once the staining with BCIP is completed, staining with 1% w/v eosin is performed to obtain visual contrast with the region of the indigo color. At the end, dehydration, clarification, and mounting are done as previously described.

Experimental Results

In order to evaluate the effect of calcium on intestinal alkaline phosphatase activity, *in situ* isolation of intestinal loop was performed in female adult 200-g IIM/Fm subline "m" rats. Animals were anesthetized with urethane (120 mg/100 g body weight) by intraperitoneal injection. Room temperature was maintained at 21 to 22 °C. Four centimeters of duodenum were isolated *in situ* by two ligatures, one of them at the level of the pylorus. An incision in the duodenal wall was performed with an iris scissors, through which a catheter was inserted. Two ml of 50 mM calcium solution was introduced through the catheter in treated rats. Control rats were subjected to the same maneuvers, and the lumen was instilled with 2 ml of saline solution without calcium.[7] After 20 minutes, specimens of the duodenum were obtained and processed as stated above. Digital analyses were performed over images captured with a camera joined to a microscope. Optical density of the indigo compound formed by enzymatic action was measured with a specific software in the brush border membrane. Figure 42.9 displays results as mean ± SEM.

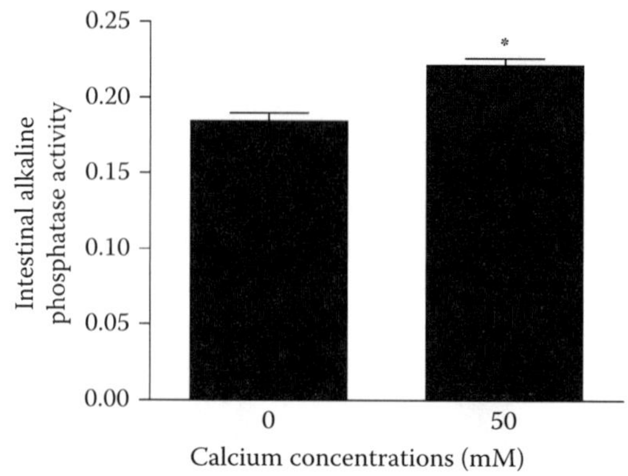

FIGURE 42.9 *In vivo* intestinal alkaline phosphatase activity. Asterisk (*) indicates significant differences with respect to controls ($p < 0.05$).

Immunohistochemistry for Rat Intestinal Alkaline Phosphatase

Immunohistochemistry is the detection of compounds with the aid of monoclonal or policlonal antibodies, which can join specifically to cellular components (antigen). Antibodies are conjugated with enzymes or fluorescent substances in order to make them visible. The technique that is described here was designed for the detection of intestinal alkaline phosphatase in the brush border of the duodenum of rats.

Adhesion of slices must be done with a one-tenth dilution of vynilic glue, and they must be kept at room temperature for 90 minutes until completely dry.[8] Once the slices have adhered to the glass, they are stored for 12 hours at 60°C. Deparaffination is performed by treatment with xylol for 30 minutes and then 2 minutes with methanol or ethylic alcohol. Slides are immersed in 2.5% w/v hydrogen peroxide 100 volumes for 30 minutes, and then they are immersed in ethylic alcohol 96° for 10 minutes and ethylic alcohol 70° for another 10 minutes (Figure 42.10).

After washing the slides three times with Tris buffer 0.05 M pH 7.6 for 5 minutes, they are treated with 1% w/v sodium citrate for 10 minutes at 100 °C in a microwave oven, which is performed in order to unblock antigens. After reaching room temperature, the slides are washed three times with Tris buffer 0.05 M, pH 7.6 for 5 minutes. The slides are treated with 1/100 dilution of primary antiintestinal alkaline phosphatase for 60 minutes. After washing the slides with Tris buffer, incubation in 1/5000 dilution of anti-IgG peroxidase conjugated secondary antibody is performed for 60 minutes. Finally, and after washing with Tris buffer, slides are treated with peroxidase substrate: 3-amine-9-ethyl carbazole (AEC)[9] or 3,3'-diaminebenzidine (DAB)[10] until observation of color. After that, washing and mounting with glycerin is done (Figure 42.11).

Experimental Results

In situ isolated intestinal loop experiments were carried out in 200 g IIM/Fm rats. The lumen was filled with 0, 10, 50, and 100 mM Ca (five repetitions). After 20 minutes, samples of mucosa were obtained. The expression of intestinal alkaline phosphatase (IAP) was measured

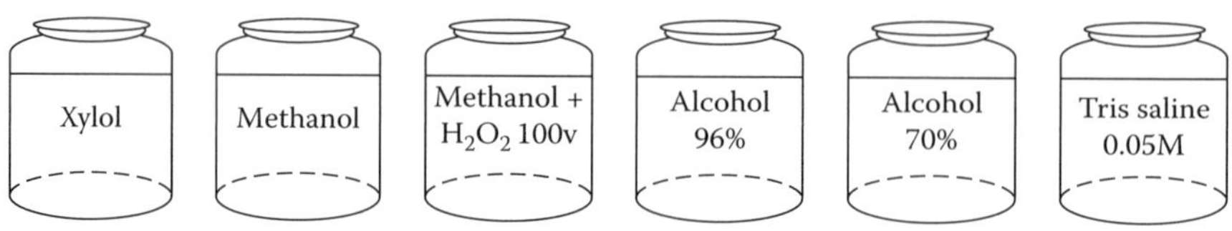

FIGURE 42.10 From left to right, solution for immunohistochemistry.

FIGURE 42.11 From left to right, solution for immunohistochemistry.

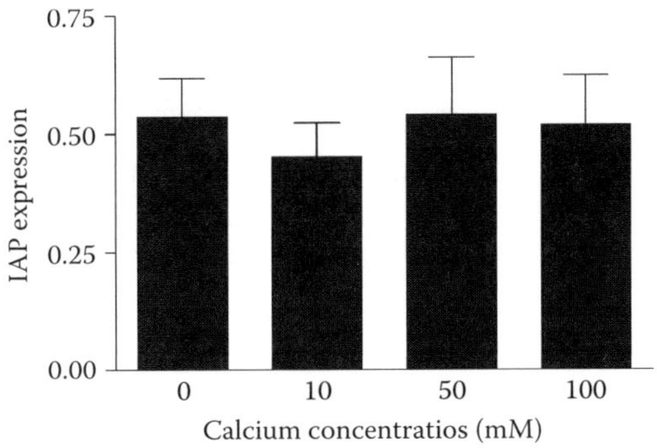

FIGURE 42.12 IAP expression in brush border membrane of duodenum is illustrated. Results are expressed as mean ± SEM.

(immunohistochemistry) with an anti-IAP primary antibody and a peroxidase-conjugated secondary antibody. Digital images were captured with a microscope and a video camera, and optical density of the compound formed by peroxidase activity was measured with a specific software. Brush border expression of IAP was not different between groups (analysis of variable (ANOVA), $p > 0.05$) (Figure 42.12).

REFERENCES

1. Preece, A. 1972. *Manual for histologic technicians*, 3rd ed. Boston: Little, Brown & Co.
2. Di Fiore, M. 1959. *Diagnóstico histológico*. 4th ed. Buenos Aires, Argentina: Editorial El Ateneo.
3. Ross, M.H., L.J. Romrell, and G.I. Kaye. 1997. *Histología. Texto y Atlas Color*, 3rd ed. Editorial Médica Panamericana.
4. Everson Pearse, A.G. 1972. *Histochemistry: Theoretical and applied*, 3rd ed. London: Churchill Livingstone.
5. Massachusetts General Hospital and Harvard Medical School, eds. 2000. *Current protocols in molecular biology*. New York: John Wiley & Sons.
6. Horwitz, J., and J. Freisler. 1970. Substrates for cytochemical demonstration of enzyme activity. V. thymidime 3´- and 5´- (5-Bromo-4-chloro-3-indolyl) phosphates. *J. Med. Chem.* 13(5):.1024–1025.
7. Brance, M.L., L.R. Brun, L. de Candia, A. Rigalli, and R.C. Puche. 2006. *In vivo* effect of calcium on brush border intestinal alkaline phosphatase in the rat. *Bone* 38(5): S16.
8. Ramos-Vara, J.A. 2005. Technical aspects of immunohistochemistry. *Vet. Pathol.* 42: 405–426.
9. Graham, R.C. Jr, U. Lundholm, and M.J. Karnovsky. 1965. Cytochemical demonstration of peroxidase activity with 3-amino-9-ethylcarbazole. *J. Histochem. Cytochem.* 13:.150–152.
10. Fahimi, H.D., and V. Herzog. 1973. A colorimetric method for measurement of the (peroxidase-mediated) oxidation of 3,3'-diaminobenzidine. *J. Histochem. Cytochem.* 21(5):.499–502.

43 Bones and Bone Tissue

Hilda Moreno, Mercedes Lombarte, and Verónica E. Di Loreto

BONES AND BONE TISSUE

Utility

The obtaining of bones from a rat is necessary when size, mass, and mineral content need to be measured. Although there are noninvasive ways to measure morphometric variables and bone mineral density, their direct measurement needs the separation of bones from soft tissues. Once this is done, the analysis of the mineral content and its composition can be completed. The method in this chapter describes the obtaining of bones; the measurement of morphometric variables, such as length, mass, and diameter; and the study of their bone mineral content. It is also necessary to obtain the bones for the study of proteins from the bone matrix, for histology and histomorphometric studies, and for biomechanical measurements. In these situations, bones are obtained exclusively with the aid of scalpels, scissors, and forceps. However, in some situations, it is necessary to obtain them after digestion of soft tissues with papain. The procedure is described in detail below.

Materials

General anesthetic
Rat board
Ring-handled preparation scissors
Ring-handled iris scissors
Cotton pliers (2)
Adson forceps
Rat-tooth forceps
10-ml syringe
Injection needle 25 × 0.8
Container for autoclave
Autoclave
Sieve
37°C drying oven
60°C drying oven
Papain
1% w/v sodium azide
Acetone
6 M hydrochloric acid solution
Paper bags
0.1-mg sensitivity scale
Porcelain crucible
550°C incinerating oven

Procedure

After euthanasia, the rat is skinned and eviscerated to obtain the carcass (skeleton plus muscles). Following this, the carcass is put in a heat-resistant glass container, covered with water, cupped and sterilized in an autoclave for 20 minutes at 120°C. Then, when the container returns to room temperature, 10,000 U papain are added to the content, close to a Bunsen burner to maintain the sterility. At this time, 5 ml of 1% w/v sodium azide is added to prevent bacterial growth. The container is maintained at 37°C for 2 to 3 days, with intermittent shaking to facilitate the action of the enzyme and the digestion of soft tissues. After obtaining the complete digestion and separation of bones from soft tissues, the container is removed from the 37°C oven and the content is passed through a sieve with holes, small enough to not allow the passage of the smallest bones of the rat. A sieve with holes no bigger that 1 mm in diameter is recommended.

Finally, the soft tissue is eliminated with a jet of tap water, and only the bones remain in the sieve. With the aid of magnifying glasses and cotton pliers, the bones are put in a container with acetone to eliminate fat. The procedure involves three passages in acetone, 2 days each. Finally, the bones are put in a paper bag in a 60°C drying oven. The weight of the bones is measured in subsequent days until a constant weight is obtained. This process can take as long as 5 days.

Precautions

Sodium azide is extremely hazardous and can be life threatening. Therefore, it must be handled carefully. Gloves need to be worn when working with the solution which cannot be discarded in the sewer after the procedure is completed. Instead, a special container in which this kind of product can be chemically inactivated should be used.

Radiological Densitometry and Morphometric Measurement

The bone mineral density can be measured by x-ray attenuation. It is useful to assess changes in density at trabecular or cortical level. The radiographs can be done with x-ray odontologic equipment.

First, an aluminum calibration stepwedge and solutions of calcium and phosphate, each with different concentration (calibration curve), are exposed simultaneously to the x-ray. The absorbance of the solutions is expressed as the function of calcium concentration. Then, the absorbance of the selected points of the aluminum stepwedge is converted in calcium concentration (g/cm^2) with the mentioned function. The calibration of the stepwedge means the solutions of calcium and phosphate do not have to be used in the future.

Then, for each x-ray film, the function of net absorbance units (average absorbance at each step of the wedge minus absorbance of background) versus g Ca/cm^2 at each step of the wedge is obtained.

The absorbance of bones are measured simultaneously with the stepwedge and with the aid of the mentioned function, the bone mineral density (g Ca/cm^2) is obtained (Figure 43.1) . Absorbance of the stepwedge and bones can be measured on digital images of the x-ray, with the appropriate software.

It is important to have the power, focus-film distance, and exposure time correctly set to obtain the best quality images. This is also true for the development and fixing of the films.

Morphometric measurements, such as length (L), width (W), and cortical thickness (C), can also be measured on the same radiograph. As cortical thickness depends on the distance from the diaphysis, a definite place must be chosen for the measurement, e.g., the middle point between diaphysis, and the measurement is done in a perpendicular direction to the longitudinal axis of the bone.

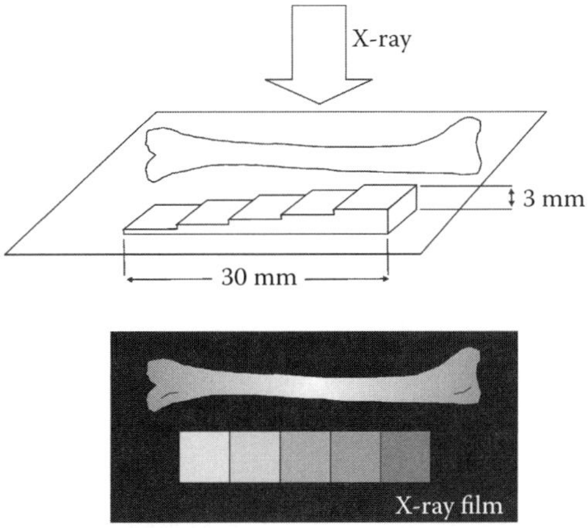

FIGURE 43.1 Radiography of the tibia and the stepwedge is displaced.

Experimental Results

The right tibia was obtained from Sprague Dawley rats (n = 8) weighing 292 ± 12 g, as described above. Length, width, and cortical thickness were measured on the x-ray. Results are shown in Table 43.1

Bone Mineral Content Analysis

Dry bones are weighed, put in a porcelain crucible, and incinerated at 550°C for 8 hours. After reaching room temperature, the ashes are weighed. The weight of organic bone matrix is obtained by subtracting the weight of the ashes from the weight of the dry bones

$$\text{Organic bone matrix} = \text{dry skeleton} - \text{ashes}$$

The percentage of bone mineral (%BM) can be obtained calculating the ratio:

$$\%\ \text{BM} = \text{weight of ashes}*100/\text{weight of dry skeleton}$$

This measurement can be considered as an index of general mineralization of bone tissue. The same information can be assessed with the ratio weight of ashes/weight of matrix. These indexes provide information about the entire skeleton or the bones under study. However, the results must be used carefully considering that bone remodeling is a process that occurs at microscopic level, and mineralized matrix co-exists with nonmineralized matrix.

The measurement of phosphorus and calcium content, as well as other ions, can be performed after dissolving the ashes in 6 M hydrochloric acid.

TABLE 43.1
Morphometric Measurement of the Right Tibia

Dry Weight (g)	Length (mm)	Total Width (mm)	Cortical Thickness (mm)
0.3312 ± 0.0072	35.30 ± 0.1103	2.537 ± 0.2356	0.7171 ± 0.1030

TABLE 43.2
Mineral Content of Femur in Control and Diabetic Rats

	Controls (n = 12)	Diabetics (n = 14)
Body weight (g)	583 ± 35	273 ± 34*
Glycemia (mg%)	106 ± 5	616 ± 116*
Ashes (mg/g dry bone)	670 ± 6	617 ± 3.1*
Matrix (mg/g dry bone)	329 ± 6	383 ± 3.1*
A/M	2.04 ± 0.05	1.61 ± 0.02*
Calcium (mg/g dry bone)	256 ± 5	261 ± 2
Phosphate (mg/g dry bone)	264 ± 15	208 ± 3*
Ca/P	0.98 ± 0.07	1.26 ± 0.008*

* Indicates significant differences with respect to controls $p < 0.05$.
Note: A/M = weight ashes/weight matrix ratio. Data are expressed as mean ± SEM.

Experimental Results

Diabetes was induced in 21-day-old male IIM/Fm rats by a single intraperitoneal injection of 24 mg of alloxan per 100 g body weight. The objective of the experiment was to study the effect of hyperglycemia and hyperphosphaturia on bone matrix and mineral composition. After 17 weeks, 14 rats with nonfasting plasma levels of glucose above 300 mg% were selected. A group of same aged and same sex rats (n = 12) were used as controls. The animals were euthanized, the right femur dissected and defatted with acetone, dried, and weighed with an analytical scale. Bone mineral content was measured as described above. The results are shown in Table 43.2.

Section XI

Miscellaneous

44 Equipment

Alfredo Rigalli

THERMOSTATIZED SURGERY RAT BOARD

The surgery rat board, especially the one used in surgeries where the rat remains under anesthetics for periods longer than 15 minutes, should have the ability of maintaining the body temperature of the animal. There are boards that contain water, but can be inconvenient, e.g., high weight and difficult maneuverability.

However, a thermostatized rat board can be constructed rather easily. It should consist of three clearly visible parts:

1. The rat board itself
2. Electronic thermostat: temperature control system
3. Heating rat board

Rat Board

It consists of a 20 × 30 cm by 1.5 to 5 mm thick stainless steel sheet with rounded edges. The rat is stuck to the sheet with sticking tape and the surgery is performed over it. This sheet is not joined together with the heating panel and this allows for easier surgery maneuver. Once the surgery is finished, the temperature of the panel can be monitored through a thermometer placed in the heating rat board, and the temperature of the animal through a rectal thermometer.

Electronic Thermostat

Materials and Electronic Components

Integrated circuit TDA 1023 (16 pins DIL)
Triac TIC 226 D or any other for 8 A and 400 V (with heat sink)
Resistor: all of ¼ watt
- R1: 22 KΩ
- R2 y R3: 4,7 KΩ
- R4: 50 KΩ
- R5 y R6: 820 Ω
- R7: 100 Ω
- R8: 56 Ω

Lineal potentiometer of 22 KΩ
Capacitors:
- C1: 470 μF 25 V electrolytic
- C2: 47 nF 400 V
- C3: 47 μF 25 V electrolytic
- C1: 330 nF 630 V
- C5: 33 nF 400 V

Thermistor: 27 KΩ to 25 °C NTC (negative temperature coefficient)

Electric terminal for seven wires
Ferric chloride
Copper board for printed circuit
Indelible felt-tip pen or nail varnish
Drill
1 to 2 mm drill bit
Soldering iron
Solder

Someone with basic knowledge of electronics can make the board and the success of the job is based on the accurate construction of the printed circuit. In Figure 44.1, the printed circuit components side is shown.

In Figure 44.2, the copper side of the printed circuit board is shown.

Precaution

To test the circuit, the board resistor of the heating rat board must be connected; otherwise, irreparable damage can be caused to the integrated circuit TDA 1023.

Heating Rat Board

The parts of the heating board are identified with the letters A, B, C, D, etc. in Figure 44.3. A represents the board itself, B and F are two aluminum boards of 2 to 3 cm thickness separated by an asbestos panel with U (C) shape, D is a board resistor, and E is an asbestos panel that isolates

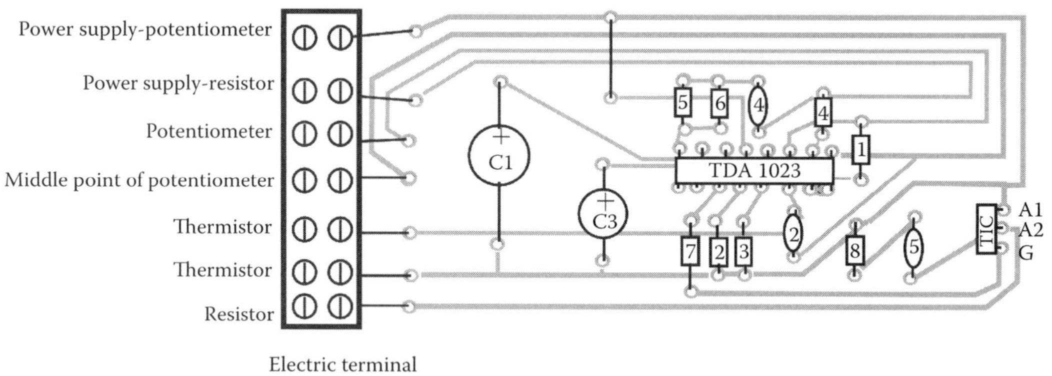

FIGURE 44.1 View of the components side of the printed circuit board.

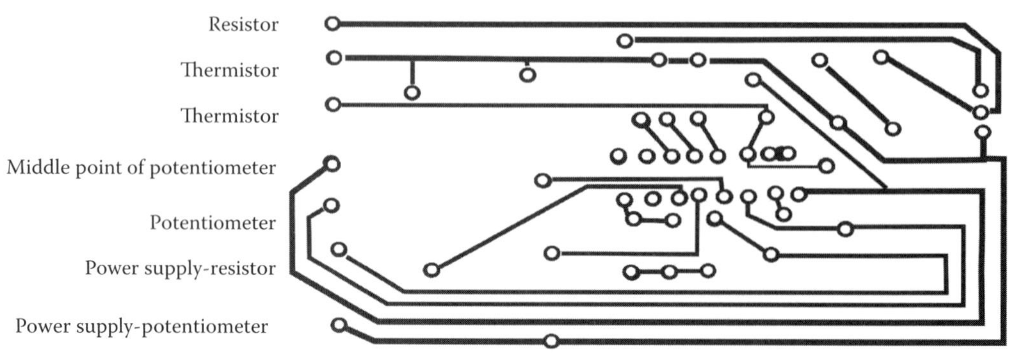

FIGURE 44.2 View of the copper side of the printed circuit board.

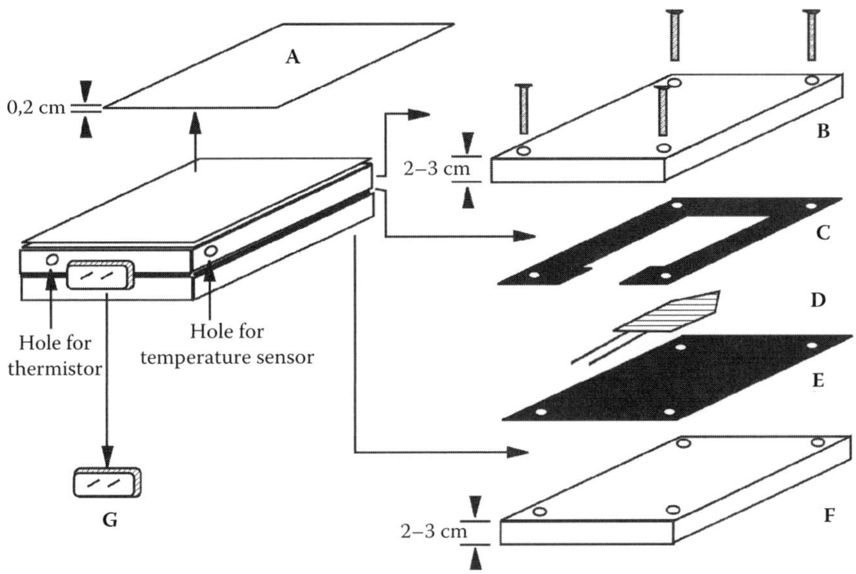

FIGURE 44.3 Joint of the heating panel.

the resistor of the aluminum board of the lower side. B, C, E, and F are joined through four three-sixteenth-inch rivets.

Board B should have two orifices, one for the thermistor that will send the signal to the thermostat, and the other where the thermometer can be inserted to monitor temperature. Both orifices should have the adequate depth for the previously mentioned sensors. Once parts B, C, E, and F are joined with the resistor inside, the terminals of the latest will be connected with the terminals of the electronic thermostat. For a better maneuverability of the equipment, it is suggested to connect the terminals of the resistor to a plug that will supply the electronic thermostat and the heating rat board independently.

HEATING LAMPS

To keep the temperature constant with or without the operating board, an incandescent or infrared lamp can be used. The major inconvenience of the lamps is that they produce more dryness in the mucus and epithelium than the thermostatized rat board. The advantages are its simplicity and low cost.

Temperature on the skin of the animal as a function of lamp-to-rat distance for different kinds of lamps is depicted in Figure 44.4.

As it can be observed in Figure 44.4, exponential equations adjust the values of temperature to lamp-to-rat distance. Using these equations, the temperature for a certain lamp and distance can be easily calculated.

DIRECT MEASUREMENT OF ARTERIAL BLOOD PRESSURE

Blood pressure can be measured by direct or indirect methods. Indirect methods involve plethysmography and direct methods require the cannulation of one artery (carotid, aorta, or femoral) and the rat must be under a general anesthetic. The materials and procedure for catheterization of the artery are described in Chapter 13.

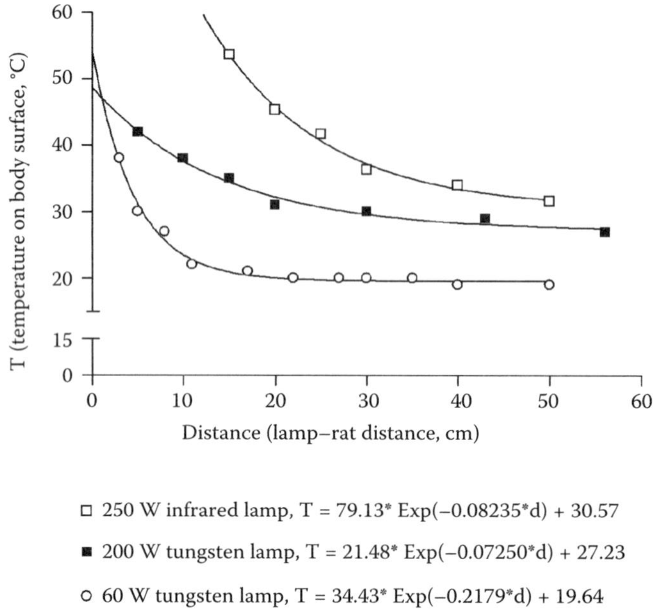

FIGURE 44.4 Temperature on the surgical surface as a function of the distance rat-to-lamp (d), and equation for the functions that fit experimental values.

The Manometer

The manometer can be easily made. It is comprised of one three-way valve (valve B), one two-way valve (valve A), one 10-ml syringe, 3-mm inner diameter polyethylene tubes, and a millimeter-graduated ruler. Mercury is needed for the construction of the device, and it should be filled with mercury maintaining the entire instrument inside a flat container with 2 to 3 cm edges, in case the mercury leaks. Precautions must be taken with the manipulation of mercury, as it spreads easily while leaking and reacts spontaneously with some metals like gold. It is important to wear gloves in order to avoid contact with skin and metallic objects.

A scheme displaying the manometer's parts is shown in Figure 44.5.

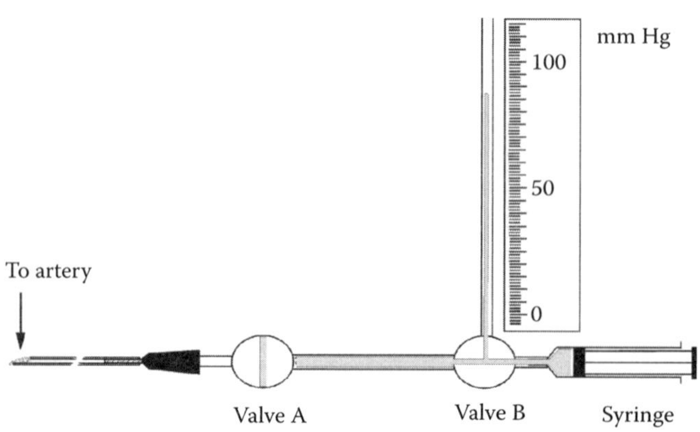

FIGURE 44.5 The relationship between components of the manometer for direct measurement of blood pressure.

The following steps must be completed carefully with the mercury container and the manometer inside the flat container.

1. With the valves A and B in the positions displayed in Figure 44.5, the syringe is filled with 10 ml of mercury and inserted in valve B (Figure 44.6a).
2. Mercury is flowed through the catheter until it reaches approximately 100 mm in the ascendant catheter, near the ruler (Figure 44.6b).
3. The ascendant catheter is turned off with valve B, maintaining open the horizontal catheter. Valve A must be open and, with the aid of the syringe, mercury is flowed until it reaches the left end of the beveled catheter to be inserted in the femoral artery (Figure 44.6c).
4. Then, the beveled end of the catheter is placed in saline solution with heparin, and the piston of the syringe is moved to flow the solution into the catheter until it reaches the left side of valve A (Figure 44.6d).
5. In the next step, the valve A must be turned off and valve B must connect the horizontal and vertical catheters (Figure 44.6e).
6. The artery is cannulated with the catheter that is at the end of valve A. The end of the beveled catheter that is in the femoral artery must be placed at the same height of the "zero" on the ruler. Valve A is opened and the mercury in the ascendant catheter should move following the pulse of blood with an oscillation of approximately 1 mm (Figure 44.7). The value of blood pressure is measured in mmHg in the ruler close to the vertical catheter.

Precautions

The beveled catheter must be no longer than 3 to 4 cm in order to avoid resistance. The system must be free of bubbles that can produce excessive inertia of the system.

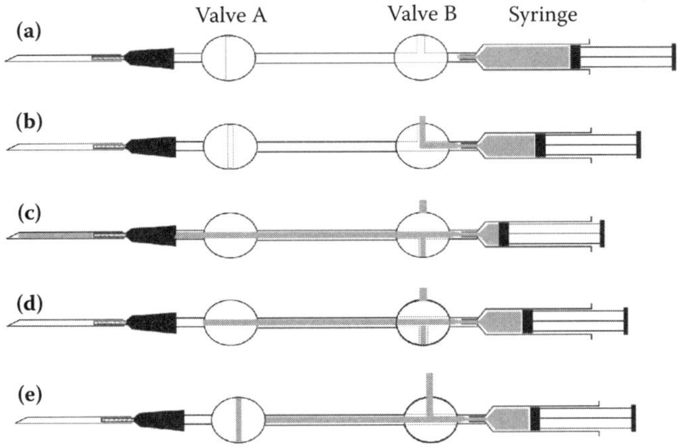

FIGURE 44.6 Movements of valves and syringe to prepare the manometer for direct measurement of blood pressure.

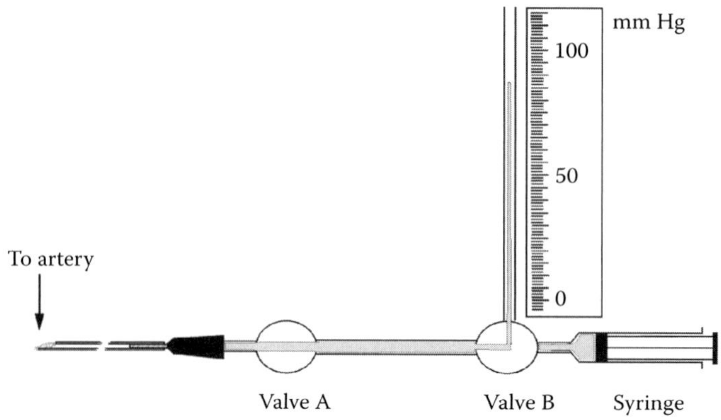

FIGURE 44.7 Position of valves while measuring blood pressure is pictured.

45 Solution Preparation

Verónica E. Di Loreto and Alfredo Rigalli

COMPOSITION OF SOLUTIONS USED IN EXPERIMENTAL MODELS

Acetic Acid 0.2% w/v

0.19 ml of glacial acetic acid + enough quantity of water to make a final volume of 100 ml.

Phosphotungstic Acid 5% w/v

5 g of phosphotungstic acid + enough quantity of water to make a final volume of 100 ml.

5-bromo-4-chromo-3-indolyl phosphate (BCIP) for Histochemistry

Stock solution: Dissolve one tablet of BCIP in 1 ml of N,N-Dimethylformamide (DMF) 100%; stable for 1 year at 4°C in caramel-colored container.

Solution to use: dilute 48 µl of BCIP solution stock in 7 ml of APSB buffer. Stable 1 hour at room temperature.

APSB Buffer

Quantities for a 100 ml preparation:

40 ml of Tris 0.25 M + 3.25 ml of NaCl 180 g/l + 10 ml $MgCl_2$ 50 mM + enough water to make a final volume of 95 ml + adjust to pH 9.5 + enough water to make a final volume of 100 ml.

With the mixture of these quantities, the final concentrations obtained are: 100 mM Tris, 100 mM NaCl, 5 mM $MgCl_2$.

Krebs–Ringer Buffer (KRB)

Composition: 142 mM NaCl, 5.5 mM KCl, 1.4 mM KH_2PO_4, 1.4 mM $MgSO_4$, 3.1 mM $CaCl_2$, gas with carbogen, and adjust to pH 7.4.

Krebs–Ringer Buffer Bicarbonate (KRBB)

Composition: 115 mM NaCl, 4.7 mM KCl, 1.2 mM KH_2PO_4, 1.2 mM $MgSO_4$, 2.56 mM $CaCl_2$, 5 mM $NaHCO_3$, 0.2 % w/v bovine serum albumin (BSA), 20 mM HEPES, pH 7.4.

Krebs–Henseleit Buffer

Composition: NaCl 118 mM, $CaCl_2$ 2 mM, KCl 5.9 mM, $MgSO_4$ 1.2 mM, $NaHCO_3$ 20 mM, and dextrose 11.1 mM.

Buffered Formaldehyde 10% w/v

100 ml formaldehyde 40% + 195 ml Na_2HPO_4 400 mM + enough water to make a final volume of 1000 ml, pH 7.4.

Ponceau-Fuchsin

Stock solution: 0.2 g Ponceau + 0.1 g acid fuchsin + 0.2 g Orange G + glacial acetic acid 0.2% + enough water to make a final volume of 30 ml.

Solution to use: take 10 ml of the stock solution and mix with 90 ml of acetic acid 0.2% w/v.

Weigert's Hematoxylin

Solution 1: 1 g Hematoxylin + ethyl alcohol + enough water to make a final volume of 100 ml.

Solution 2: 4 ml of ferric chloride 29% w/v + 95 ml of distilled water + 1 ml of hydrochloric acid 36 wt%, density 1.18 g/ml.

Mix 1 and 2 in equal parts immediately before using.

Heparin for Intravenous Injection

Dilute 0.03 ml of heparin 5000 IU/ml in 0.5 ml of saline solution or KRB. The injection of this quantity produces the heparinization of the animal for 4 hours.

NaCl 180 g/l

180 g NaCl + enough water to make a final volume of 1000 ml. Stable at room temperature for indefinite time.

Saline Solution (NaCl 9 g/l or 0.9% w/v NaCl)

50 ml of NaCl 180 g/l + enough water to make a final volume of 1000 ml. Stable for 1 week at room temperature; it may present bacterial or fungal contamination in a short term period. It must be prepared every week.

Urethan 120 mg/ml

12 g of urethan + enough water to make a final volume of 95 ml. Adjust pH 7 to 7.4. Complete with water up to 100 ml. Stable at room temperature. Handle with care: carcinogenic and eyes, skin, and mucosa irritant.

Light Green

0.1 g light green + acetic acid 0.2% w/v + enough water to make a final volume of 100 ml.

Mayer's Albumin

Filtrated egg white and glycerine in equal parts + two or three thymol crystals added.

Phosphate Buffered Saline (PBS)

200 ml Na_2HPO_4 400 mM + 100 ml KH_2PO_4 200 mM + 32.5 ml NaCl 180 g/l + enough water to make a final volume of 1000 ml, pH = 7.5.

46 Normal Values in IIm/Fm Rat Subline "M"

Verónica E. Di Loreto and Alfredo Rigalli

The following tables show the values of some variables of the rat.

	Body Weight (g)			Water Intake (ml/day)			Food Intake (g)			Urinary Volume (ml/day)		
Day	mean	SD	n	mean	SD	n	mean	SD	n	mean	SD	n
21	43	8	30	7	3	11	9	5	11	0.9	0.9	9
22	43	4	11	10	3	10	9	4	12	0.3	0.2	5
23	45	5	12	9	2	10	8	3	12	N/A	N/A	N/A
24	46	5	12	10	2	12	10	4	12	N/A	N/A	N/A
26	50	4	12	11	1	12	9	3	12	N/A	N/A	N/A
27	57	5	12	12	3	12	15	4	12	N/A	N/A	N/A
28	60	6	10	13	3	11	12	3	12	N/A	N/A	N/A
29	64	6	12	14	3	12	11	2	12	1.7	0.8	12
30	75	11	20	13	3	12	12	3	12	2.5	1.2	12
31	74	7	12	16	3	11	19	7	12	1.3	1.0	12
33	80	7	12	N/A	N/A	N/A	N/A	N/A	N/A	N/A	N/A	N/A
34	92	10	12	16	3	12	15	2	12	2.9	1.1	12
35	106	16	20	18	4	12	21	6	12	0.6	0.4	12
36	101	7	12	16	2	12	19	1	12	2.0	0.9	12
37	107	8	12	15	2	12	17	1	12	2.6	1.1	12
38	111	8	12	16	3	11	18	4	12	2.0	1.2	12
39	116	9	12	18	5	12	N/A	N/A	N/A	N/A	N/A	N/A
40	N/A	N/A	N/A	N/A	N/A	N/A	N/A	N/A	N/A	5.5	2.7	12
41	115	9	12	22	4	12	17	5	12	4.6	2.1	12
42	122	11	11	21	3	12	19	2	12	5.0	1.6	12
43	130	23	8	N/A	N/A	N/A	N/A	N/A	N/A	N/A	N/A	N/A
49	166	19	22	22	4	6	17	4	6	5.7	2.2	6
50	N/A	N/A	N/A	24	6	6	21	2	6	5.5	1.4	7
51	157	9	6	12	6	6				9.6	3.6	6
52	136	10	6	30	9	6	12	5	6	13.8	5.1	5
55	143	9	6	N/A	N/A	N/A	N/A	N/A	N/A	N/A	N/A	N/A
56	167	28	14	30	5	6	21	4	6	9.8	3.7	5
57	154	8	6	24	5	6	20	2	6	11.6	5.2	6

—continued

	Body Weight (g)			Water Intake (ml/day)			Food Intake (g)			Urinary Volume (ml/day)		
Day	mean	SD	n	mean	SD	n	mean	SD	n	mean	SD	n
59	205	23	7	N/A	N/A	N/A	N/A	N/A	N/A	N/A	N/A	N/A
64	205	31	8	N/A	N/A	N/A	N/A	N/A	N/A	N/A	N/A	N/A
69	211	26	8	N/A	N/A	N/A	N/A	N/A	N/A	N/A	N/A	N/A
71	193	10	6	21	3	6	21	2	6	7.6	2.2	6
72	196	9	6	22	4	6	22	1	6	7.4	2.8	6
73	198	9	6	21	6	6	21	1	6	6.1	2.3	6
75	201	10	6	26	9	6	21	4	6	N/A	N/A	N/A
77	203	15	6	25	7	5	26	1	6	11.8	7.5	6
78	208	11	6	30	16	6	21	2	6	9.9	5.4	6
79	205	13	6	28	15	6	24	1	6	11.5	6.9	6
80	207	12	6	28	11	6	23	2	6	11.1	6.7	6
81	207	13	6	29	8	6	N/A	N/A	N/A	10.7	6.2	6
84	211	14	6	N/A	N/A	N/A	23	2	6	8.0	2.0	6
85	215	13	6	33	12	6	22	2	6	11.0	6.8	6
86	216	13	6	25	4	6	22	1	6	6.5	2.7	6
88	218	14	6	N/A	N/A	N/A	N/A	N/A	N/A	N/A	N/A	N/A
89	N/A	N/A	N/A	N/A	N/A	N/A	N/A	N/A	N/A	6.1	2.7	6
90	218	13	7	N/A	N/A	N/A	17	3	6	6.1	2.7	6
91	227	23	12	26	11	6	22	2	6	6.0	4.5	6
92	225	15	6	33	18	6	22	4	6	12.3	12.2	6
93	227	17	6	17	5	6	18	1	6	5.1	2.0	6
94	226	12	6	25	5	6	20	2	6	7.9	4.3	6
99	228	12	6	N/A	N/A	N/A	N/A	N/A	N/A	N/A	N/A	N/A
100	N/A	N/A	N/A	31	13	6	24	2	6	6.4	3.9	6
101	N/A	N/A	N/A	21	4	6	22	3	6	6.6	2.8	6
104	N/A	N/A	N/A	28	12	5	21	1	6	7.2	5.4	5
106	230	12	6	31	14	6	23	4	6	7.0	4.3	6
107	234	10	6	N/A	N/A	N/A	24	3	6	N/A	N/A	N/A
108	235	14	6	N/A	N/A	N/A	23	4	6	N/A	N/A	N/A
112	237	14	6	N/A	N/A	N/A	N/A	N/A	N/A	N/A	N/A	N/A
114	231	16	6	N/A	N/A	N/A	16	3	6	N/A	N/A	N/A

Note: SD = standard deviation, n = number of rats, N/A = not applicable.

	Phosphaturia (mg/day)			Nonfasted Glycemia (g/l)			Phosphatemia (mg/dl)		
Day	**mean**	**SD**	**n**	**mean**	**SD**	**n**	**mean**	**SD**	**n**
21	1.3	1.2	2	1.2	0.1	12	9.9	1.4	7
29	8.2	3.3	2	1.2	0.2	12	N/A	N/A	N/A
30	6.3	1.4	2	N/A	N/A	N/A	10.5	1.2	11
37	10.4	5.5	2	1.4	0.2	12	8.1	1.0	10
42	11.1	7.6	2	1.5	0.2	11	9.3	2.1	12
50	13.3	0.1	2	N/A	N/A	N/A	7.9	0.0	1
52	22.7	1.8	2	N/A	N/A	N/A	8.2	0.6	5
71	17.0	0.4	2	1.5	0.1	6	7.8	2.0	6
77	2.7	0.4	2	N/A	N/A	N/A	N/A	N/A	N/A
80	16.2	4.2	2	N/A	N/A	N/A	N/A	N/A	N/A
84	16.9	1.9	2	N/A	N/A	N/A	N/A	N/A	N/A

Note: SD = standard deviation, n = number of rats, N/A = not applicable.

Index

A

B

R

S

T